Pocket Book of English Grammar for Engineers and Scientists

McGraw-Hill continues to bring you the *BEST* (**B**asic **E**ngineering **S**eries and **T**ools) approach to introductory engineering education:

Bertoline, *Introduction to Graphic Communications for Engineers,* 3/e ISBN 0073048364

Chapman, *Fortran 90/95 for Scientists and Engineers,* 2/e ISBN 0072922389

Donaldson, *The Engineering Student Survival Guide,* 3/e ISBN 0073019259

Eide/Jenison/Northup, *Introduction to Engineering Design and Problem Solving,* 2/e ISBN 0072402210

Eisenberg, *A Beginner's Guide to Technical Communication* ISBN 0070920451

Finkelstein, *Pocket Book of Technical Writing for Engineers and Scientists,* 2/e ISBN 0072976837

Finkelstein, *Pocket Book of English Grammar for Engineers and Scientists* ISBN 007352946X

Gottfried, *Spreadsheet Tools for Engineers using Excel* ISBN 0072480688

Palm, *Introduction to MatLab 7 for Engineers,* ISBN 0072922427

Pritchard, Mathcad: *A Tool for Engineering Problem Solving* ISBN 0070121893

Schinzinger/Martin, *Introduction to Engineering Ethics* ISBN 0072339594

Smith, *Teamwork and Project Management,* 2/e ISBN 0072922303

Tan/D'Orazio, *C Programming for Engineering and Computer Science* ISBN 0079136788

Additional Titles of Interest:

Andersen, *Just Enough Unix,* 5/e ISBN 0072952970

Eide/Jenison/Mashaw/Northup, *Engineering Fundamentals and Problem Solving,* 4/e ISBN 0072430273

Holtzapple/Reece, *Foundations of Engineering,* 2/e ISBN 0072480823

Holtzapple/Reece, *Concepts in Engineering,* ISBN 0073011770

Martin/Schinzinger, *Ethics in Engineering,* 4/e ISBN 0072831154

Pocket Book of English Grammar for Engineers and Scientists

Leo Finkelstein, Jr.
Wright State University

Boston Burr Ridge, IL Dubuque, IA Madison, WI New York
San Francisco St. Louis Bangkok Bogotá Caracas Kuala Lumpur
Lisbon London Madrid Mexico City Milan Montreal New Delhi
Santiago Seoul Singapore Sydney Taipei Toronto

MER
428.20245 FIN
7 day

Higher Education

POCKET BOOK OF ENGLISH GRAMMAR FOR ENGINEERS AND SCIENTISTS

Published by McGraw-Hill, a business unit of The McGraw-Hill Companies, Inc., 1221 Avenue of the Americas, New York, NY 10020.

1 2 3 4 5 6 7 8 9 0 DOC/DOC 0 9 8 7 6 5

ISBN 0–07–352946–X

Publisher: *Suzanne Jeans*
Senior Sponsoring Editor: *Michael S. Hackett*
Senior Developmental Editor: *Michelle L. Flomenhoft*
Executive Marketing Manager: *Michael Weitz*
Senior Project Manager: *Sheila M. Frank*
Senior Production Supervisor: *Sherry L. Kane*
Senior Coordinator of Freelance Design: *Michelle D. Whitaker*
Cover Designer: *Kelly Fassbinder / Imagine Design Studio*
(USE) Cover Images: *Flowchart by Leo Finkelstein; background image ©PhotoDisc, Business Today 2, Vol. 78*
Senior Photo Research Coordinator: *Lori Hancock*
Compositor: *Lachina Publishing Services*
Typeface: *10/12 Century Schoolbook*
Printer: *R. R. Donnelley Crawfordsville, IN*

Library of Congress Cataloging-in-Publication Data

Finkelstein, Leo, 1946–
    Pocket book of English grammar for engineers and scientists / Leo Finkelstein, Jr. — 1st ed.
      p. cm. (McGraw-Hill engineering BEST series)
    Includes index.
    ISBN 0–07–352946–X
    1. English language—Grammar—Handbooks, manuals, etc. 2. English language—Technical English—Handbooks, manuals, etc. 3. Technical writing—Handbooks, manuals, etc. I. Title. II. Series.

PE1475.F47   2006
428.2—dc22
                                        2004065608
                                        CIP

www.mhhe.com

For Phyllis and Stephen

About the Author

Leo Finkelstein, Jr., received a bachelor's degree from the University of North Carolina at Chapel Hill in 1968, a master's from the University of Tennessee at Knoxville in 1969, and a Ph.D. from Rensselaer Polytechnic Institute at Troy, New York, in 1978. He is currently lecturer and director of technical communication for the College of Engineering and Computer Science, Wright State University, Dayton, Ohio. During his 20-year Air Force career, he was involved in a wide range of technical, academic, and professional activities. He directed the technical writing program at the U.S. Air Force Academy, which, at the time, was the largest technical writing program in the country. He also wrote, produced, and directed many films, mostly on technical subjects, while in Southern California. During the Vietnam War, he commanded a combat photographic unit in Southeast Asia, where he developed innovative, aerial photographic techniques and processes. In addition, his military service includes experience in space and logistics systems, as well as high-level, corporate communications. He holds FCC commercial and amateur radio licenses, has a black belt in Taekwondo, and is an avid user of all types of technology. He loves gadgets!

Contents

Preface

I wrote this book to give busy, time-challenged engineering and science students an abbreviated but comprehensive guide to the most important fundamentals of written American English. My goal was to develop a practical reference that effectively and efficiently helps these students understand, recognize, and fix common grammatical errors. While specifically geared to the needs of engineering and science students in general, this guide provides a simple, user-friendly approach that is also appropriate for almost any student, and is particularly appropriate for English as a Second Language (ESL) students. The emphasis is on simplicity, clarity, and understandability.

This book also makes use of several innovative approaches geared specifically to the needs of engineering and science students at both the undergraduate and graduate levels. Flowcharts, tables, and diagrammed examples are provided throughout to help students effectively apply the rules of grammar to their specific requirements. Also, with few exceptions, the examples in the book relate in some way to engineering and science.

Purpose and Approach

Organization   This book is divided into the following sections:

- *Introduction* (Chapter 1). This chapter provides an overview of the eight parts of speech, reviews basic information related to ESL needs, and deals with fundamental sentence structure and punctuation.
- *Parts of Speech* (Chapters 2-9). These eight chapters discuss each of the eight parts of speech by chapter. Chapters provide detailed discussions, examples, data tables, and flow-charts, as appropriate.
- *Punctuation* (Chapter 10). This chapter covers the 14 common marks of punctuation in English and shows how, when, and where to use each mark. This chapter also provides detailed examples, a quick reference table, and flowcharts.
- *Final Thoughts* (Chapter 11). This is a brief conclusion to the first ten chapters.
- *Glossary* (Chapter 12). This comprehensive, alphabetically organized section provides a listing of terms related to grammar and usage. This section also provides definitions, examples, explanations, and cross-references for virtually all of these terms.
- *Index.* This section contains an alphabetized list of names, subjects, and topics, and is designed to provide quick, random access to specific information in the book.

Acknowledgments

I express my sincere appreciation to all my many friends and colleagues who did so much to help me with this book. Specifically, I thank Phyllis A. Finkelstein, my best friend and wife of 35 years, for her nonstop encouragement and patience. She has also proven to be an extraordinary proofreader. I thank my old Air Force buddy, William G. Dwyer, an English professor and Shakespeare scholar at the Air Force Academy "back in the old days," and now a firmly entrenched "big thinker" in the corporate world. I appreciate all the time Bill has spent philosophizing with me about the rules of grammar and reviewing my manuscript. I thank my friend and colleague, Thomas A. Sudkamp, professor of computer science and engineering and successful book author, for the many hours he has spent discussing with me the finer points of grammar— from restrictive/nonrestrictive clauses to the subjunctive mood. Throughout the development of this book, I have relied heavily on Tom as a sounding board and "sanity check."

I also appreciate the superb inputs received from the many reviewers for this book, including: Sharon Ahlers, Cornell University; Jeanne Linsdell, San Jose State University; Darryl Morrell, Arizona State University; John Brocato, Mississippi State University; Cesar Luongo, Florida State University; John Gershenson,

Michigan Technological University; Marjorie Rush Hovde, Indiana University-Purdue University Indianapolis; Donna T. Matsumoto, Leeward Community College-University of Hawaii; and, Tim Jordanides, California State University, Long Beach.

Finally, my sincere appreciation to my sponsoring editor, Carlise Stembridge, who came up with the idea for this book, and who has provided infinite patience, strong leadership, and superb guidance and management for this project.

Leo Finkelstein, Jr.
College of Engineering and Computer Science
Wright State University
Dayton, Ohio

01
00001

Introduction

Welcome to the *Pocket Book of English Grammar for Engineers and Scientists,* a codified guide to the basic rules of written American English. This book is designed to provide you with the information you need to quickly and effectively understand and use the major concepts of English grammar in your technical reports and presentations.

1.1 Importance of Grammar

Grammar is nothing more than a collection of the rules that we use for assembling words so that, together, they make sense and convey meaning. Making sense and conveying meaning *precisely* is the primary function of technical writing. Incorrect grammar can affect the sense of what you are trying to say by making it more difficult to understand, or even worse, by changing your meaning altogether. Poor grammar poses significant risks for the credibility and effectiveness of any technical document.

Poor grammar also says something personally about you and the quality of your work. Grammatical errors in a technical report can compromise not only the value of the report, but also the perception of you as a professional scientist or engineer. Poor grammar on your part can communicate to the reader that you are not the "brightest person in the room," or that you lack the required education and professional attention to detail. True or false, right or wrong, fair or unfair, grammatical errors can seriously undermine your credibility.

In a technical document, you are judged to a great extent based on the document's quality. By the time someone reads what you have written, it is too late. You are not there to defend yourself, you are not available to explain what you really meant to say, and you have no opportunity to fix your mistakes.

This book is designed to help you address grammar issues up front, while you are still writing and editing the document—and before the reader tries to understand your meaning or makes a judgment regarding the quality of your work or your thinking. You will probably find this book's greatest value to be in the editing process, because editing is the key to good, effective technical writing. You should always plan on your first draft being lousy because first drafts are notoriously deficient. The secret to good writing is editing and rewriting! Remember, the final draft is usually the only one that counts.

**1.2
Parts of Speech**

The English language has eight parts of speech: *nouns, pronouns, adjectives, verbs, adverbs, prepositions, conjunctions,* and *interjections.* These are the fundamental building blocks of sentences, and as such, they are critical for understanding how English works (Figure 1.1). Each of the following eight chapters is organized around and focuses on a specific part of speech.

Figure 1.1
Parts of speech.

English is an imprecise, inconsistent, and illogical language that can be frustrating and difficult to use—and that's just for those who are fluent in it! Nonnative speakers who use English as a Second Language (ESL) often find the process of writing and speaking the language to be much more challenging. The author can empathize just a bit with ESL writers because he remembers one of his first English teachers, a tyrannical, Catholic nun who terrorized him during his grade school years. Whenever he would misspell a word, she would berate him for not looking it up in the dictionary. His defense was based on the principle of practicality (and desperation), arguing that he would have consulted a dictionary if the word had only looked wrong to him, but it looked right—and if he were to check all the words that look right, along with all the words that look wrong, he would be checking all the words all the time. Unfortunately, that was exactly what the teacher had in mind.

**1.3
Grammar and
English as a
Second
Language**

The language challenge the author experienced then is somewhat analogous to what ESL writers face today—theirs is the same type of problem on a much larger scale, but hopefully without the tyrannical teacher. Native English speakers do not consciously apply rules of grammar while writing. They go along until something intuitively "feels" wrong. Maybe a misspelled word catches their eye, or a verb's tense sounds strange. For many native speakers of English, potential problems often just stand out and demand attention. Nonnative users of English frequently do not have this advantage. For many ESL writers, errors do not just stand out, and there is no intuitive indication or "feeling" that something might be wrong.

Modern word processors now make "checking every word" both practical and routine. Although far from perfect, computerized proofing tools can and do alert us to a wide range of potential problems—from misspellings to verb agreement errors. In fact, armed with modern proofing tools and a basic knowledge of how the English language works, ESL writers, along with everyone else, can become more effective technical writers.

**1.4
Sentence
Structure**

Because sentences are so important to understanding the parts of speech and how they work, it might be useful to briefly review the basics of sentence structure before going any further.

1.4.1 Simple Sentence

In English, a *simple sentence* is the basic, independent unit of expression. All simple sentences are composed of two major grammatical divisions: the *subject*, which is a noun or pronoun about which something is asked or asserted; and the *predicate*, which is doing the asking or asserting. The part of the predicate that provides the action is called the *verb*. For example, in the simple sentence that follows, *processor* is the <u>subject</u> and *outperforms* is the <u>verb</u>.

The new *processor outperforms* its predecessor.

To show this schematically, we can diagram this sentence as shown in Figure 1.2.

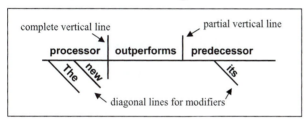

Figure 1.2
Sentence diagram #1.

The sentence's subject is to the left of the complete vertical line, while everything to the right of this line is the *predicate*. The verb is separated from the rest of the predicate by a partial vertical line. Below the horizontal line are diagonal lines containing modifiers for both the subject and object. Figure 1.3 is a more completely labeled diagram of this sentence and may help further clarify this structure.

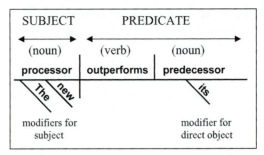

Figure 1.3
Sentence diagram #2.

Sentences can be *simple, compound,* or *complex.* The sentence above is a *simple sentence* because it follows the pattern of SUBJECT–PREDICATE. It is also called a *main clause* or *independent clause* because it has everything it needs to stand alone as a sentence.

1.4.2 Compound Sentence

A *compound sentence* is essentially two *main clauses* joined together with a *coordinating conjunction* to make a longer sentence (for more on *conjunctions,* see Chapter 8).

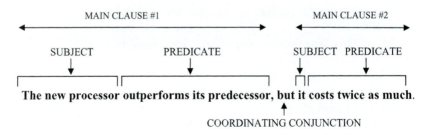

1.4.3 Complex Sentence

A *complex sentence* is one that contains a *main clause* and a *subordinate clause*. The *main clause* (also called an *independent clause*) can stand alone, while the *subordinate clause* (also called a *dependent clause*) cannot stand by itself—it depends on the presence of the main clause.

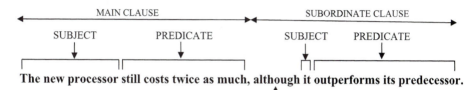

In the sentence above, the subordinating conjunction *although* makes the second clause *it outperforms its predecessor* dependent on the first clause, which is the main clause.

1.4.4 Compound-Complex Sentence

The final type of sentence is the *compound-complex*. Here a compound sentence, formed from two main clauses, has one or more of these main clauses joined with one or more subordinate clauses. In the example below, the sentence has two main clauses (*I know*, and *I do not care*). Notice that the first main clause is joined to a subordinate clause (*that the new process still costs twice as much*).

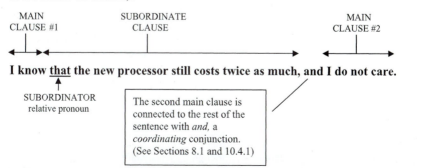

The second example (below) of a compound-complex sentence has a similar pattern of two main clauses, *the new processor costs twice as much* and *I do not care*, and one subordinate clause, *although I will regret it.* Here, the subordinator is a *subordinating conjunction* instead of a *relative pronoun.*

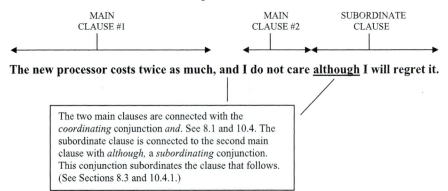

| MAIN CLAUSE #1 | MAIN CLAUSE #2 | SUBORDINATE CLAUSE |

The new processor costs twice as much, and I do not care <u>although</u> I will regret it.

The two main clauses are connected with the *coordinating* conjunction *and.* See 8.1 and 10.4. The subordinate clause is connected to the second main clause with *although,* a *subordinating* conjunction. This conjunction subordinates the clause that follows. (See Sections 8.3 and 10.4.1.)

Two different types of subordinators are used in these examples: a *relative pronoun* (Table 1.1) and a *subordinating conjunction* (Table 1.2). Notice that adding a subordinator changes a *main clause*, which is also a complete sentence, into a *subordinate clause*, which is no longer a complete sentence. Remove the subordinator, and the subordinate clause becomes an independent, main clause again—one that can stand alone as a complete sentence.

Table 1.1   Common relative pronouns

that	which	whoever	whomever
what	who	whom	whose

Table 1.2   Common subordinating conjunctions

after	if	unless	wherever
although	once	until	while
as	since	when	
because	that	whenever	
before	though	where	

1.4.5 Basic Punctuation for Sentences

Although the punctuation used in a sentence will be varied and specific to the style, structure, and application, the basic punctuation of any sentence involves a concluding or *terminal* mark at the end (*period, question mark, exclamation point*), and under certain conditions, commas where two or more clauses are joined. For a more detailed discussion of punctuation, see Chapter 10.

Table 1.3   Basic sentence punctuation

SIMPLE	Add a concluding, terminal mark at the end of the sentence.
COMPOUND	Add a comma before the coordinating conjunction and a terminal mark at the end of the sentence.
COMPLEX COMPOUND-COMPLEX	Add a comma before a *coordinating conjunction* that joins main clauses. Add a comma before a *relative pronoun* only if its dependent clause is not essential to the meaning of the sentence (*nonrestrictive**) (Chapter 3). Add a terminal mark of punctuation at the end of the sentence.

*A *nonrestrictive* element (noun, pronoun, phrase, clause) in a sentence is one that is not essential to the meaning of the sentence. A *restrictive* element is one that is essential. Normally only nonrestrictive elements are set off with commas (see 4.4, 10.4.4).

CHAPTER TWO

Nouns

Nouns are the words used to name *persons* or *animals* (Albert Einstein, Marie Curie, Coco the ape), *places* (London, Mt. McKinley, Rensselaer Polytechnic Institute), *things* (transducers, networks, wave guides), *concepts* (relativity, file path, dynamics), *qualities* (intelligence, significance, homeostasis), or *actions* (insertion, dissipation, dilation). When using nouns, you should be aware of several important variables: *number, type, class, case,* and *gender.*

**2.1
Definition and
Functions**

Number is the simple concept that refers to whether the noun includes one (*singular*), two or more (*plural*), or both. For example, if I have one rock, a quartz specimen, then the noun *specimen* is <u>singular</u>. If I have several rocks, five quartz specimens, then the noun *specimens* is <u>plural</u>. If I have a single word that refers to all my rock specimens, maybe my rock collection, then *collection* would be singular since it represents a *single* collectivity of all the rocks.*

**2.2
Number**

---

*Collective nouns can be confusing, especially for ESL students whose initial contact with English was with British English. British English often treats collective nouns as plural. For example, the collective nouns *team* and *company* in British English are often considered to be plural, while in American English, the same nouns would be singular. See Glossary, "English: British vs. American."

Normally we make a singular noun plural by adding *-s* or *-es* to the end. So one molecule becomes two or more *molecules*, one *radar* becomes five *radars*, and one *standard* becomes two or more *standards*. Easy enough, right? Well, not so fast. Unfortunately, the English language is full of illogical, mind-twisting exceptions.

2.2.1 Irregular Nouns

Many nouns do not follow the *-s* or *-es* rule. So if we have a *mouse* here and a *mouse* there, together we do not have two *mouses*. Whether we are talking about computer pointing devices or small rodents, the noun *mouse* is irregular and its plural is mice. So we have two *mice*.

The same is true of the so-called *mass nouns* or *uncountable nouns* that are not quantified by how many, but rather, by how much—or are not quantified at all. "Sand at the seashore" usually refers to the overall mass of the sand, not to how many individual grains are present. When we buy sand in the hardware store, we buy it by its mass (or, more precisely, by its weight). We do not use the individual unit of grains (Figure 2.1).

Figure 2.1
Mass/uncountable nouns. Mass/uncountable nouns like *sand* are both singular and plural for obvious reasons.

Table 2.1 Examples of mass/uncountable nouns

compassion	fortitude	peace	sugar
courage	gravel	sand	tranquility
flour	love	strength	water

There are also many other exceptions—in English, nothing is easy. For example, words that end in -s may be either singular or plural depending on the context of their use. In fact, sometimes the same word is treated differently depending on the context and its meaning.

Economics 400 **is a difficult class.**

> Here the word *economics* is singular because it is used to describe an academic course. Consequently, the verb is singular.

The *economics* **of going to school are also difficult.**

> Here the word *economics* is plural because it is used to describe financial considerations. Consequently, the verb is plural.

Another exception to the rule involves words that are both singular and plural.

We have 25 *aircraft* **on the flight line, and my** *aircraft* **is at the far end.**

> As used here, the noun *aircraft* is plural (not aircrafts).

> As used here, the noun *aircraft* is singular.

About 100 *deer* **are in the herd and one** *deer* **is in my backyard.**

> The noun *deer* is plural (not deers).

> The noun *deer* is singular.

We have the *means* to connect to the Internet.

> The noun *means* is also both singular and plural. In this sentence, *means* probably includes an Internet service provider, a cable modem, a router, and a computer. It is clearly plural, but it would never be written as *means<u>es</u>*.

Table 2.2 lists examples of irregular nouns and their plurals.

Table 2.2   Examples of irregular nouns and their plurals

<u>A</u>	<u>E</u>	<u>K</u>
addendum, addenda	echo, echoes	knife, knives
alumnus, alumni	elf, elves	
analysis, analyses	ellipsis, ellipses	<u>L</u>
antenna, antennae/	embargo, embargoes	leaf, leaves
antennas	emphasis, emphases	life, lives
apparatus, apparatuses	erratum, errata	loaf, loaves
appendix, appendices		louse, lice
axis, axes	<u>F</u>	
	fish, fish	<u>M</u>
<u>B</u>	focus, focuses	man, men
bacterium, bacteria	foot, feet	matrix, matrices
basis, bases	formula, formulas	means, means
bureau, bureaus	frequency, frequencies	medium, media
	fungus, fungi	memorandum, memoranda
<u>C</u>		millennium, millennia
cactus, cacti	<u>G</u>	moose, moose
calf, calves	genus, genera	mosquito, mosquitoes
child, children	goose, geese	mouse, mice
corps, corps		
crisis, crises	<u>H</u>	<u>N</u>
criterion, criteria	half, halves	nebula, nebulae
curriculum, curricula	hero, heroes	neurosis, neuroses
	hypothesis, hypotheses	nucleus, nuclei
<u>D</u>		
datum, data	<u>I</u>	<u>O</u>
deer, deer	index, indices	oasis, oases
diagnosis, diagnoses		ovum, ova
		ox, oxen

P	species, species	V
paralysis, paralyses	stimulus, stimuli	vertebra, vertebrae
parenthesis, parentheses	stratum, strata	veto, vetoes
person, people	syllabus, syllabi	vita, vitae
phenomenon, phenomena	symposium, symposia	volcano, volcanoes
potato, potatoes	synopsis, synopses	
	synthesis, syntheses	W
R		watch, watches
radius, radii	T	wife, wives
	tableau, tableaux/tableaus	wolf, wolves
S	teeth, teeth	woman, women
scarf, scarves	that, those	
scissors, scissors	thesis, theses	Z
self, selves	thief, thieves	zero, zeroes
series, series	tomato, tomatoes	
sheep, sheep	torpedo, torpedoes	
shelf, shelves		

Nouns are either *proper* or *common*. *Proper nouns* name a specific person, animal, place, thing, concept, or quality, while *common nouns* are nonspecific (Table 2.3). The first letter of any proper noun is almost always capitalized no matter where it occurs in a sentence. Exceptions include certain business names and trademarks that may not begin with a capital letter as part of their visual identity—e.g., eTrade Financial® or eBay®.

2.3 Type

Table 2.3 Examples of proper and common nouns

Proper	Common
Albert Einstein (the most famous physicist)	physicist (any physicist)
Coco (a famous primate)	ape (any primate)
Princeton (a famous university)	university (any university)
Earth (the third planet from the Sun)	earth (dirt under our feet)
Theory of Relativity (revolutionary theory)	scientific theory (any theory in science)
Dyson's Sphere (place to live in space)	sphere (anything shaped like a ball)

2.4
Case

Case is the category of a noun or pronoun that indicates its function in a sentence (Table 2.4). In English, nouns and pronouns can have three cases: *nominative* or *subjective, objective,* and *possessive. Nominative* and *subjective* mean the same thing.

Table 2.4 Functions of nouns for each case

Case	Noun's function
Nominative/Subjective	Subject of a verb or verb's complement* (does the action)
Objective	Object (directly or indirectly receives the action)
Possessive	Shows possession, ownership, control, or custody

*As used here, the verb's *complement* includes any words that complete the sense of the verb.

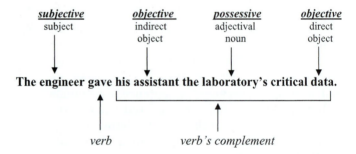

In this sentence, *engineer* is in the subjective or nominative case because it functions as the subject of the sentence. The noun *assistant* is in the objective case because it serves as the indirect object by telling us to whom or for whom the action is done. The noun *laboratory* is in the possessive case because it shows possession of the *critical data.* Finally, the noun *data* is in the objective case because it serves as the direct object and receives the verb's action.

Table 2.5 Making nouns possessive

Number	Punctuation	Example
Singular possessive noun that does not end in *s*	Add an *apostrophe* + *s* to the end of the word	computer's cost (only one computer)
Singular possessive noun that ends in *s*	Add just an *apostrophe*, or add *apostrophe* + *s* to the end of the word (whatever looks better)	Phyllis' Website Phyllis's Website
Plural possessive nouns	Add just an *apostrophe* to the end of the word if it ends in *s*. Add *apostrophe* + *s* if the word does not end in *s*.	computers' cost (plural word ends in *s*) vertebrae's location (plural word does not end in *s*)

2.4.1 Gerunds and Case

As mentioned earlier, a word's specific function in a sentence determines what part of speech it is. *Verbals* (gerunds, adjectives, and participles) are words normally used as *verbs* that are functioning as *nouns, adjectives,* or *adverbs* in a particular sentence. The focus here is on the *gerund*, a verb ending in *–ing* that is functioning as a noun. We will look at participles and infinitives later when we deal with adjectives (Chapter 4) and adverbs (Chapter 6). For now, think "noun" and "gerund" and consider the following sentences:

Calibrating vectorscopes can be time-consuming.

> The verb *to calibrate* is a gerund acting as a noun. Because it serves as the subject of the sentence, the gerund *calibrating* is in the subjective case.

I dislike *calibrating* vectorscopes because it is very time-consuming.

> As before, the verb *to calibrate* is a gerund acting as a noun. In this case, however, it serves as the direct object of the verb *dislike;* consequently, it is in the objective case.

When dealing with gerunds and case, one other important rule to remember is that the case of the gerund's "subject" is always <u>possessive</u>. The *subject of a gerund* is really just about anything that immediately precedes the gerund in a sentence. In the following sentence, the gerund *calibrating* is preceded with a possessive pronoun *his*, which in turn acts as the subject of the gerund *calibrating*.

Subject of the gerund

Gerund

His calibrating **the vectorscope required more time than we had budgeted.**

In the following example, the gerund *failing* has the personal pronoun *him* immediately preceding it. This is incorrect because *him* is in the objective case. The possessive form of the pronoun *his* should be used here.

Subject of the gerund

his Gerund

The supervisor attended the meeting to discuss him failing to do his work on time.

In the following sentence, the subject of the gerund is the noun *technician*, which is in the possessive case because it immediately precedes the gerund *calibrating* in the sentence.

Subject of the gerund

Gerund

The *technician's calibrating* **of the vectorscope required more time than we had budgeted.**

English uses three genders: *male, female,* and
neuter. Nouns may or may not be gender specific, but either way, they do not require special
treatment based on their gender. However, as we
will see in the next section, certain pronouns
must agree in gender and number with the nouns
they replace (their *antecedents*).

**2.5
Gender**

Many nouns that exist in the English language
were once widely used but are now considered
offensive and disrespectful. Most relate to race,
ethnicity, gender, and age. For example, in the
1950's, it was common to refer to attractive
women as "broads," and African Americans as
"Colored People." ESL writers, especially those
new to the language, are at particular risk of
inadvertently using such words after picking
them up from documents, motion pictures, and
television shows produced at a time when such
language was more acceptable. All writers, especially ESL writers, must be extremely careful in
this regard, because such use today could easily
compromise the credibility and effectiveness of
a technical document or briefing for a given
audience.

**2.6
Offensive Nouns**

Table 2.6 Common offensive nouns and acceptable alternatives

Offensive	Acceptable	Offensive	Acceptable
actress	actor	little woman	wife
bellboy	bellhop	maid	housekeeper
broad	woman	mankind	humankind
chairman	chairperson	mongolism	Down Syndrome
Christian name	given name	Negro	African American*
colored	African American*	old maid	single woman
Eskimo	Inuit, Native Alaskan	policeman	police officer
fireman	firefighter	postman	letter carrier
forefather	ancestor	spokesman	spokesperson
girl (adult)	woman	steward/stewardess	flight attendant
Girl Friday	assistant	sweetie (w/ strangers)	woman
housewife	homemaker	weatherman	meteorologist
layman	nonspecialist	workman	worker

*Black is also an acceptable alternative.

2.7
Appositives
An appositive is a noun or noun phrase placed beside another noun to describe, explain, or identify that noun. In effect, an appositive renames the noun it is placed beside.

Pahoehoe lava, *a textured or smooth formation*, occurs extensively on the Kilauea volcano in Hawaii.

> Here, *a textured or smooth formation* is an appositive. It is in apposition with *Pahoehoe lava.*

Note that in this sentence, commas set off the appositive because it is nonrestrictive. In other words, while the appositive clarifies what Pahoehoe lava is, the appositive is not essential to the meaning of this sentence.

2.8
Noun Clauses
A *noun clause* is a subordinate clause used in its entirety as a noun in a sentence. Look at the following sentences, each of which uses a noun clause in a different way:

Whatever you receive for this idea has to be more than it is worth.

> This noun clause is being used as the **subject** of the sentence.

You will be paid *whatever you are worth.*

> This noun clause is being used as the **direct object** of the sentence.

I will pay *whoever shows up* more than I'm paying you.

> This noun clause is being used as the **indirect object** of the sentence.

I will pay a bonus <u>to</u> *whoever is the most productive.*

> This noun clause is being used as the **object of the preposition** *to*. Notice that, as in the prior example, the subject of this clause *whoever* is in the subjective case, even though the entire clause is in the objective case.

03
00011

Pronouns

A *pronoun* is a generic word that takes the place of a noun in a sentence and functions in the sentence exactly as the noun it replaces.

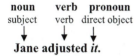

noun	**verb**	**noun**
subject	verb	direct object

Jane adjusted the *signal generator*.

The direct object *signal generator* can be replaced with the pronoun *it*.

noun	**verb**	**pronoun**
subject	verb	direct object

Jane adjusted *it*.

In this sentence, *it* functions as the direct object, just as *signal generator* did.

3.1.1 Antecedents and Ambiguity

Most pronouns must have a specific word or group of words to which they clearly refer. This referent, called the pronoun's *antecedent*, usually occurs before the pronoun. By context or position, it should be obvious to the reader exactly which word or words serve as the antecedent.

In the following sentence, the pronoun *this* is ambiguous.

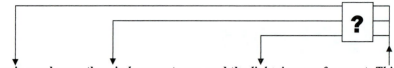

The *rain* was heavy, the *winds* were strong, and the *lightning* was frequent. *This* concerned the forecasters greatly.

Exactly what concerned the forecasters? Heavy rain? Strong wind? Frequent lightning? Maybe they were concerned about a combination of some or all of these phenomena? There is no way to know by reading the sentence. The pronoun *this* could refer to one, two, or three of the possible referents, or any combination of some or all of them. As used in the example, the pronoun *this* is ambiguous because it has no clear antecedent. An easy way to fix this problem is to add the appropriate noun after *this*.

storm

The *rain* was heavy, the *winds* were strong, and the *lightning* was frequent. This ∧ concerned the forecasters greatly.

The noun *storm* has been added after the pronoun *this*, and the sentence now clearly means the totality of the storm, not just some of its phenomena.

The next sentence provides another example of ambiguous pronoun reference. Here, in the second main clause, we are not sure what *it* refers to.

The *thunderstorm* intensified along the *dry line*; clearly, *it* was a serious threat.

Punctuation note: A semicolon joins the two main clauses and acts as a "soft" coordinating conjunction. Semicolons are often used to join two closely related clauses. (See 10.13.)

Perhaps the antecedent is the *thunderstorm*—certainly with all that rain, wind, and lightning, the thunderstorm is a threat. However, another interpretation is possible: that *it* refers to the *dry line*, an active boundary between hot dry and warm moist air that has the potential for generating many severe thunderstorms.

An easy way to fix this problem is to replace the pronoun *it* with the appropriate noun or nouns, as shown in the following example.

The thunderstorm intensified along the dry line; 〈 **the *storm* was a serious threat.**

OR

the *active boundary* was a serious threat.

3.1.2 Antecedents and Agreement

Pronouns agree with their antecedents in *person*, *number*, and *gender*. In other words, if the antecedent is in the first person, the pronoun must be in first person; if the antecedent is plural, then the pronoun must be plural; and, if the antecedent is masculine, the pronoun must be masculine. Note: *indefinite pronouns* do not have antecedents (see 3.2.4).

3.1.2.1 Agreement in Person

Person refers to the identity of the subject; in other words, is the subject doing the speaking, being spoken to, or being spoken about?

- *First Person* is the person doing the speaking.
- *Second Person* is the person being spoken to.
- *Third Person* is the person being spoken about.

In English, *pronouns* and *verbs* are the only parts of speech where *person* has a grammatical significance—so pay attention to the antecedent when selecting the form of the pronoun you are using. A pronoun must be in the same person as its antecedent. Errors in agreement involving person normally do not result in a grammatically incorrect sentence, but they always result in a sentence that has a different meaning from what was intended. The following sentence shows agreement in the third person between the subject *man* and the pronoun *his*. In this sentence the meaning is clear: this guy did not like his assignment.

The *man* did not like *his* assignment.

In the following sentence, the pronoun no longer agrees in person. The antecedent *man* is in the third person, while the pronoun *your* is in the second person.

The *man* did not like *your* assignment.

While this sentence is grammatically correct, the meaning of the sentence is changed. The reader would be forced to search for, or at least try to assume or imagine, a second person antecedent; otherwise, the sentence would not make sense.

3.1.2.2 Agreement in Number

Number refers to how many things the antecedent represents.

- *Singular* is used when the antecedent represents only one thing.
- *Plural* is used when the antecedent represents more than one thing.

The following example uses the same sentence as before, but this time to show agreement in number. The pronoun *his* agrees in number with its antecedent *man*. Both are singular.

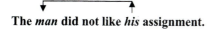

The *man* did not like *his* assignment.

In the following sentence, the pronoun *their* agrees with the subject *man*, in person (both are in the third person), but no longer agrees in number because *their* is plural and *man* is singular.

The *man* did not like *their* assignment.

As was true with the previous agreement-in-number error, agreement errors in person are not necessarily grammatically incorrect, but they always confuse and sometimes change the meaning of a sentence. In this example, the man is no longer concerned about <u>his</u> assignment, but rather, he is concerned about some group's assignment.

3.1.2.3 Agreement in Gender

Gender refers to whether the antecedent is *masculine, feminine,* or *neuter*. Pronouns must agree with their antecedents in gender.

- *Masculine* refers to antecedents that are *male*.
- *Feminine* refers to antecedents that are *female*.
- *Neuter* refers to antecedents that are neither male nor female.

For example, the two sentences below reference a programmer named Mary who is, in fact, using her own laptop computer.

Since the subject is female, the pronoun *her* also must be female.

***Mary* used *her* laptop computer.**

In the next sentence, a gender discrepancy exists between the female subject *Mary* and the referencing male pronoun *his*.

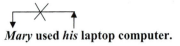

Mary used *his* laptop computer.

The sense of the sentence has been changed significantly because of this simple agreement error. We do not know the antecedent of the pronoun *his*, but we do know it is not *Mary*. The antecedent has to be male, unless, of course, Mary is a guy with a very strange name. A reader might well search the material preceding this sentence to find a male antecedent. That would result in the worst type of error in technical writing—one that conveys unintended and incorrect information.

3.2
Types
of Pronouns

Many types of pronouns exist. While each pronoun has the common function of substituting for a noun, each type has characteristics that make it unique. The types of pronouns, based on subfunction, include *personal, reflexive, intensive, indefinite, relative, interrogative, demonstrative,* and *reciprocal.*

3.2.1 Personal Pronouns

Personal pronouns are used to substitute for specific people or things. In the following sentence, the pronoun *she* substitutes for the noun *physicist*, a person; while the pronoun *their* substitutes for the noun *lasers,* a thing.

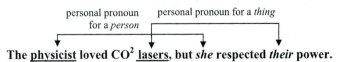

The <u>physicist</u> loved CO^2 <u>lasers,</u> but *she* respected *their* power.

Personal pronouns, unlike nouns, usually have different forms for different cases (see Table 3.1). For example, the personal pronoun for the first person singular could be *I*, *my*, or *me* depending on whether it is in the subjective, possessive, or objective case. The function of the pronoun in a sentence (acts as the subject, object, or shows possession) determines what case (subjective, objective, possessive) is required. The following example uses the first person singular pronoun in all three cases.

subject	shows possession	object of preposition
subjective	possessive	objective
↓	↓	↓

I used *my* computer, which had been provided for *me*.

The following compound sentence contains three personal pronouns, each in a different case. Two of the pronouns, *his* and *he*, have *scientist* as their antecedent, while the pronoun *it* refers to the *lab*. The pronoun *his* shows possession and is in the possessive case. The pronoun *he* serves as the subject of the second main clause and is in the subjective case. The pronoun *it* is the object of the preposition *in* and is in the objective case.

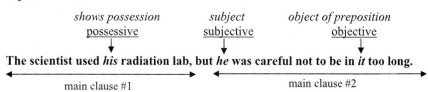

Personal pronouns also agree with their antecedents in gender. This requirement can be confusing because gender only applies to third person, singular, personal pronouns. All other personal pronouns in English are not gender specific.

Table 3.1 provides a matrix of all personal pronouns arranged by case, number, person, and gender.

Table 3.1 Personal pronouns by case, number, person, and gender

	Subjective		Objective		Possessive	
Number	**Singular**	**Plural**	**Singular**	**Plural**	**Singular**	**Plural**
1st person	I	we	me	us	my/mine	our/ours
2nd person	you	you	you	you	your/yours	your/yours
3rd person	he (**m**) she (**f**) it (**n**)	they	him (**m**) her (**f**) it (**n**)	them	his (**m**) her/hers (**f**) its (**n**)	their/theirs

m=male   f=female   n=neuter

3.2.1.1 Gender Specificity and Sexist Usage

As noted above, in English, the third person singular is gender specific. For many years, the masculine form of the third person singular was used generically to refer to both male and female antecedents. For example:

Everyone in the laboratory must do *his* part.

The problem with this sentence is that not everyone in the laboratory is male, and many see the generic use of male pronouns to refer to everyone as offensive and demeaning to women. If all employees of the lab were male, then the use of *his* would be appropriate (although the hiring practices of the lab might be questionable). Additionally, if we add *her* to the equation (*his* or *her* part), then in what order should they appear? Also, do we alternate between *his and her* and *her and his*, and if so, exactly how often? As you can see, this approach can be problematic.

Fortunately, the English language provides an easier solution for most situations. Remember, the third person plural is not gender specific. So to avoid any of these problems, simply write in the *third person plural*. In the sentence below, make the antecedent plural and change the pronoun to *their*.

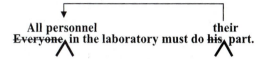

All personnel   **their**
~~Everyone~~ in the laboratory must do ~~his~~ part.

3.2.2 Reflexive Pronouns

Reflexive pronouns are personal pronouns used as an object or complement of a verb. (See Table 3.2.) They are unique in that they not only receive or elaborate on the verb's action from the subject, but also reflect the action back to the subject. *Reflexive pronouns* are easy to spot because they have *-self* or *-selves* in their spelling. The sentence below contains an example of both a singular and plural reflexive pronoun.

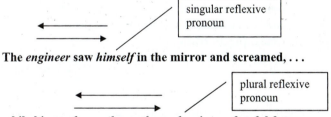

singular reflexive pronoun

The *engineer* saw *himself* in the mirror and screamed, . . .

plural reflexive pronoun

while his *employees* threw *themselves* into a fearful frenzy.

Here, the reflexive pronoun *himself* receives the verb's action as the direct object, and then reflects the action back to the subject *engineer*. The engineer actually receives his own action! The same thing is true of the *employees*, who as the subject, threw *themselves*.

Table 3.2 Common reflexive pronouns

myself	herself	ourselves
yourself	itself	yourselves
himself	oneself	themselves

3.2.3 Intensive Pronouns

Intensive pronouns look like reflexive pronouns (Table 3.2), but they differ in their function. *Intensive* pronouns are used to emphasize another word in a sentence, while *reflexive* pronouns reflect the action back from the verb to the subject. In the following example, the intensive pronoun *herself* intensifies and emphasizes the antecedent *boss*.

The boss *herself* praised my C programming knowledge and accuracy.

3.2.4 Indefinite Pronouns

As mentioned earlier, English is a language of exceptions to rules, and indefinite pronouns fit into this category. An *indefinite pronoun* is a pronoun that does not require an antecedent. In other words, it can substitute for a noun in a sentence without specifying a specific person, animal, place, thing, etc.

indefinite pronoun

If we conduct the experiment without controls, *nothing* will come of it.

In this sentence, the pronoun *nothing* does not have an antecedent but can still stand alone as a noun. Table 3.3 provides examples of indefinite pronouns.

Table 3.3 Examples of indefinite pronouns

all	both	few	nobody	several
another	each	many	none	some
any	either	more	no one	somebody
anybody	everybody	most	nothing	someone
anyone	everyone	much	one	something
anything	everything	neither	other	such

3.2.5 Relative Pronouns

Relative pronouns are noun substitutes that are used to introduce subordinate clauses. For that reason, relative pronouns are often known as *subordinating pronouns* or *subordinate-clause markers* (see Table 3.4).

Table 3.4 Common relative pronouns

Most common	Less common	
who	what	whichever
that	whoever	whatever
which	whomever	

3.2.5.1 That, Which, *and* Who

That and *which* are relative pronouns used to introduce subordinate clauses when the antecedent is not a person. *That* is used to introduce restrictive clauses, while *which* is used for nonrestrictive clauses. *Who* can be used to introduce either a restrictive or nonrestrictive subordinate clause when the antecedent is human. Punctuation for relative pronouns depends on whether they introduce a *restrictive* or *nonrestrictive* clause.

A *restrictive clause* is a clause that is essential to the meaning of the main clause, while a *nonrestrictive clause* is not. In the following example, the pronoun *that* introduces a subordinate clause that specifies the primary requirement for the computer.

main clause   restrictive subordinate clause

We are looking for a good computer *that* can handle full-motion, video editing.

No comma before *that* because
the clause being introduced is
restrictive

The requirement for the computer to *handle full-motion, video editing* is essential for the meaning of this sentence, and the subordinate clause is therefore *restrictive.* Notice that the relative pronoun *that* is used <u>without</u> a comma because the subordinate clause is restrictive.

In the next example, the subordinate clause tells us nothing essential about the main clause, and is therefore *nonrestrictive.* The nonrestrictive clause is introduced *with a comma* and the relative pronoun *which.**

main clause   restrictive subordinate clause

We are shopping for a new computer, <u>which might cause us to be late for dinner.</u>

A comma is required to set off
this nonrestrictive clause.

---

*In Standard American English, the distinction between *that* and *which* is slowly eroding. Increasingly, either *that* or *which* is acceptable in either restrictive or nonrestrictive applications. However, there is no guarantee that the person evaluating your writing will be so "linguistically liberal." To play it safe, especially in critical documents for which the audience is unknown, adhere to the traditional distinctions.

In the sentence below, the relative pronoun *who* is used to introduce a restrictive clause because its antecedent is a person. The clause *who just walked through the door* provides essential information and is restrictive and not set off with commas.

main clause   restrictive clause   main clause (continued)

◄——► ◄————————————————————► ◄——————————————————————►
The man *who* just walked through the door is our chief database administrator.

```
Commas are not required to
set off this restrictive clause.
```

The following example demonstrates a nonrestrictive use of the pronoun *who*.

main clause nonrestrictive clause   main clause (continued)

◄——► ◄——————————►◄——————————————————————►
The man, *who* lives in town, is our chief database administrator.

```
Commas are required to set off this non-
restrictive clause.
```

3.2.6 Interrogative Pronouns

Interrogative pronouns are simply pronouns that are used to ask questions. They frequently come at the beginning of a sentence.

Who is the chief database administrator?
Which way do I go to get there?

In this example, *who* and *which* are acting as interrogative pronouns (Table 3.5).

Table 3.5 Common interrogative pronouns

who	whom
whose	which
what	

3.2.7 Demonstrative Pronouns

Demonstrative pronouns are pronouns that point out something in the sentence.

That **sort algorithm is much less efficient for ordering large lists.**

Demonstrative pronoun. Since it modifies a noun, it also functions as an adjective.

3.2.8 Reciprocal Pronouns

Reciprocal pronouns occur in pairs and reinforce the relationship between two separate antecedents. They are unique in that both of the antecedents are referents for both of the pronouns.

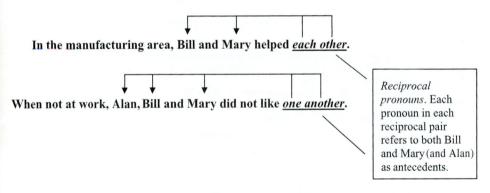

In the manufacturing area, Bill and Mary helped *each other*.

When not at work, Alan, Bill and Mary did not like *one another*.

Reciprocal pronouns. Each pronoun in each reciprocal pair refers to both Bill and Mary (and Alan) as antecedents.

3.2.9 Multi-Use Pronouns

There is one more thing you should keep in mind as you look through the lists of different types of pronouns provided in this chapter and elsewhere. The fact that a word is listed as one type of pronoun does not mean it cannot be used as another type. For example, *who, which*, and *what* are frequently used as both *relative* and *interrogative* pronouns, while *that* can be used as both a *relative* and *demonstrative* pronoun.

04

00100

Adjectives

An *adjective* is the part of speech we use to modify, describe, or limit a noun or pronoun. That seems simple enough: *fat* head, *high* building, *strong* wind, *easy* language, *ugly* dog, or *fast* car. Adjectives add details about the nouns or pronouns they modify and usually precede those nouns or pronouns. Additionally, English has no plural form for adjectives; they are always singular. So adjectives really are simple compared to some other parts of speech, but they are <u>not</u> that simple. After all, this is the English language!

4.1.1 Multifunction Adjectives

Many words we normally think of as nouns, pronouns, verbs, and adverbs are also used as adjectives—sometimes functioning simultaneously as two different parts of speech. It is like getting two parts of speech for the price of one.

For example, if a word modifies a noun, then it is functioning as an *adjective*. If that same word replaces a noun, then it is functioning as a *pronoun*. Or if the word names a person, animal, place, thing, etc., then it is functioning as a

**4.1
Definition
and Functions**

noun. In the sentence below, *his* is a personal pronoun used as an adjective modifying the noun *restaurant.*

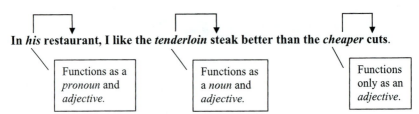

In *his* restaurant, I like the *tenderloin* steak better than the *cheaper* cuts.

| Functions as a *pronoun* and *adjective.* | Functions as a *noun* and *adjective.* | Functions only as an *adjective.* |

The word *his* has functional qualities of both a <u>pronoun</u> and an <u>adjective</u>. *His* obviously replaces a noun (a male antecedent in some prior sentence), which is clearly a pronoun function; and, at the same time, *his* modifies the noun *restaurant,* which is clearly an adjective function. Similarly, *tenderloin* names a very tasty and expensive part of a cow, so it functions as a *noun*; and it also modifies *steak* by telling us that the steak is tasty and expensive, so it functions as an *adjective* as well. That is why such dual-function adjectives are sometimes called *adjectival nouns* or *adjectival pronouns.* The adjective *cheaper* is just a plain, ordinary adjective. It functions as a single part of speech in this sentence.

4.1.2 Verbals

Verbals are nouns, adjectives, and pronouns that have the characteristics of verbs. Or to look at it from a different perspective, they are *verbs* used as nouns, adjectives, and pronouns. When verbs are used as nouns, we call them *gerunds* (see 2.4.1, Gerunds and Case). When verbs are used as adjectives, we call them *participles.* In the following example, the verb *to interest* is being used as an adjective. In both sentences, the participle form of the verb *to interest* modifies the noun *professor*. Notice the difference in meaning between the *-ing* and *-ed* forms of the participle. The *-ing* form has an external effect, while the *-ed* form is affected by external sources. You will find this difference to be true of many participles ending in *-ing* and *-ed*.

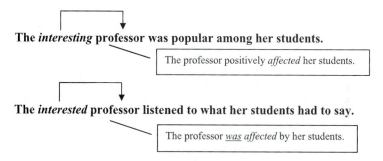

The *interesting* professor was popular among her students.

> The professor positively *affected* her students.

The *interested* professor listened to what her students had to say.

> The professor <u>was</u> *affected* by her students.

4.1.3 Misplaced or Dangling Modifier

A *misplaced* or *dangling modifier* (frequently called a *dangling participle*) is normally a participle phrase that, because of its physical placement in a sentence, does not relate clearly to the word or words it is supposed to modify. In many cases, by seeming to modify something else in the sentence, it totally changes the meaning of the sentence.

Zapped by an ungrounded electrode, the technician helped the unconscious visitor.

> Because of its placement at the front of the sentence, this participle phrase seems to modify *technician,* when in fact it should modify *visitor.*

To fix a dangling modifier, simply move it closer to the word it should modify.

The technician helped the unconscious visitor *zapped by an ungrounded electrode.*

After being moved to the end of the sentence next to *visitor*, this participle phrase clearly modifies *visitor*.

**4.2
Classes
of Adjectives**

The most common classes of adjectives are *descriptive, limiting, proper, predicate*, and *compound*. These classifications describe how the adjective modifies a noun or pronoun. In addition to these common classes, many special case adjectives exist. These include *articles* and other *determiners*, as well as complete clauses that function as adjectives in a sentence (*adjectival clauses*).

4.2.1 Descriptive Adjectives

A *descriptive adjective* simply describes the attributes of the noun it modifies. In the following example, the adjectives describe the color and brightness of the sky, the excellence of the gardens, and the poor quality of the food.

The *sunny* **sky and the** *magnificent* **gardens could not compensate for the** *lousy* **food.**

4.2.2 Limiting Adjectives

A *limiting adjective* limits the noun in some way. In the sentence below, the adjective *high* modifies the noun *frequencies*. In effect, the adjective limits *frequencies* to a specific range in the electromagnetic spectrum, since, by convention, *high frequencies (HF)* refer to those frequencies between 3 and 30 megahertz.

The RF detector functioned efficiently at *high* frequencies.

4.2.3 Proper Adjectives

A *proper adjective* is actually a proper noun used as adjectival noun. It effectively changes the sense of the noun it modifies from a common noun to a proper noun. In the following example, *Unix* changes the sense from *any* operating system to *the* Unix operating system.

The *Unix* operating system can be difficult to learn.

Proper adjectives are usually capitalized in a sentence, but only if the proper noun from which they are derived is also normally capitalized. For example, some proper nouns, especially distinctive trademarks like eTrade Financial® or eBay®, may not capitalize the first letter while capitalizing other letters. In this case, the proper adjective must match precisely the proper noun.

4.2.4 Compound Adjectives

A *compound adjective* is one made up of two or more parts, each of which describes or limits the noun in some way. In the following example, *thick, asphalt* is the compound adjective. *Thick* describes how deep the surface is, while *asphalt* describes the material from which the surface is made.

The *thick, asphalt* surface should last five years or more.

Generally, one follows a specific sequence when compounding adjectives. In the example, _thick, asphalt_ surface would be much harder to read and decipher if the adjectives were reversed to read, _asphalt, thick_ surface. To prevent this problem, we normally sequence compound adjectives starting with the determiner, which is usually an article, and ending with the noun itself. Typically, the sequence of adjectives is as follows: _determiner*, opinion, shape/size, age, color, origin, material,_ and _adjectival noun,_ followed by the noun being modified. Table 4.1 shows this sequence with examples of each category. In the above example, shape/size (_thick_) should come before material (_asphalt_).

Table 4.1 Sequencing of compound adjectives

Determiner	Opinion	Shape & size	Age	Color	Origin	Material	Adj. noun	Noun
The			new		American	titanium		fighter
A	cheap	19 inch		RGB			CRT	display
Many	good	2 gigabyte	used	gray	Korean		storage	devices

Of course, you do not want to sequence together as many adjectives as presented in Table 4.1, especially in the table's bottom row.

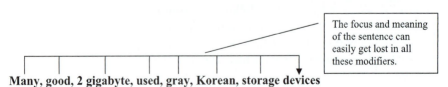

The focus and meaning of the sentence can easily get lost in all these modifiers.

Many, good, 2 gigabyte, used, gray, Korean, storage devices

---

*A _determiner_ is a special kind of adjective that signals the approach of a noun (see 4.3).

A good rule of thumb is not to sequence more than three adjectives before a noun. If more limitations or descriptions are necessary, try including them in a clause or a phrase after the noun, as shown in the following example.

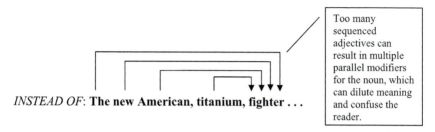

INSTEAD OF: **The new American, titanium, fighter ...**

Too many sequenced adjectives can result in multiple parallel modifiers for the noun, which can dilute meaning and confuse the reader.

TRY: **The new American fighter, made of titanium, ...**

You can also add more limitations or descriptions in a follow-on sentence.

4.2.5 Predicate Adjectives

Predicate adjectives are adjectives used with a *linking verb*. A linking verb is either a form of the verb *to be* or a *sense verb* (smell, taste, feel, see, hear, etc.). You can think of these verbs as acting like an *equal sign* in the sentence. They equate the subject to what follows the verb, and in the process, describe or limit the subject.

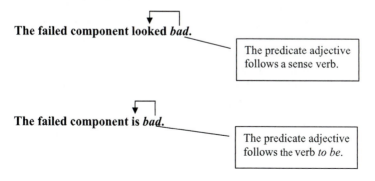

The failed component looked *bad.*

The predicate adjective follows a sense verb.

The failed component is *bad.*

The predicate adjective follows the verb *to be.*

By the way, if a *noun* follows a linking verb, it is called a *predicate nominative* or *predicate noun* and is in the *subjective* case—<u>not</u> the *objective* (see 5.5.1.1).

4.2.6 Absolute Adjectives

An *absolute adjective* is a modifier (like *infinite* or *impossible*) that turns the noun it modifies into a specific state that is not modifiable in terms of degree or intensity. The noun is absolute and cannot be changed. In the sentence below, *dead* is dead, and *perfect* is perfect.

Dead men tell no tales and *perfect* people make no mistakes.

One cannot be partially dead, really dead, somewhat perfect, or absolutely perfect. When one is dead or perfect, the result is absolute. Therefore, avoid qualifying absolute adjectives by writing expressions like "really dead" or "absolutely perfect." You can still modify absolute adjectives as long as you do not qualify them. Expressions like "nearly dead" or "almost perfect" are acceptable because the absolute state of being dead or perfect has not yet been reached.

4.3 Articles and Other Determiners

Determiners are the single-word adjectives that precede a noun and signal its approach. Determiners include both articles and adjectival pronouns. Table 4.2 provides some of the commonly used determiners.

Table 4.2 Common articles and other determiners

Articles:	a	an	the					
Other determiners:	my	their	these	this	whose	those	one	any
	our	your	his	her	its	no	some	

Generally, the rules for articles are simple: The definite article *the* modifies one specific person, place, thing, etc.

The **satellite has experienced some orbital decay.**

The indefinite articles, *a* and *an,* modify a nonspecific person, place, thing, etc. Use *a* before a noun that begins with a consonant and *an* before a noun that begins with a vowel sound (not necessarily a vowel).

A **satellite could experience orbital decay under** *a* **number of conditions.**

An **orbital decay could affect some satellites.**

ESL students often have a difficult time with articles because of the different ways articles are used from one language to another. To make things easier when using English, Figure 4.1 provides a flowchart that can help identify what article, if any, should be used.

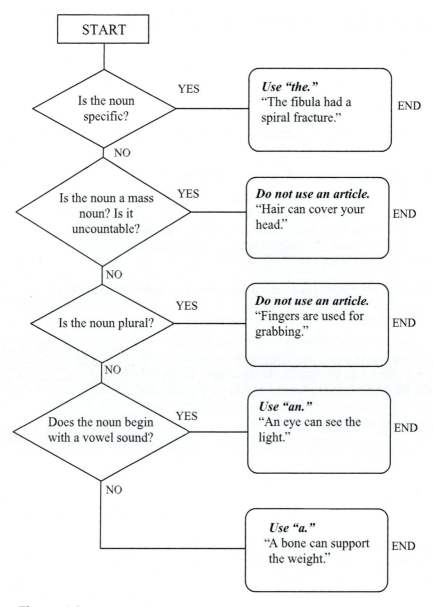

Figure 4.1
Articles—when and what to use.

Adjectival clauses are subordinate clauses that modify nouns or pronouns. In other words, the entire clause, including a subject and predicate, acts as an adjective in the sentence. Adjectival clauses are introduced with relative pronouns, most often *that, which,* or *who* (see Chapter 3, Pronouns). These clauses <u>are not</u> set off with commas when they are *restrictive,* meaning the clause is essential to the meaning or sense of the sentence; and they <u>are</u> set off with commas when the clause is *nonrestrictive,* meaning they are not essential to the meaning or sense of the sentence.

4.4 Adjectival Clauses

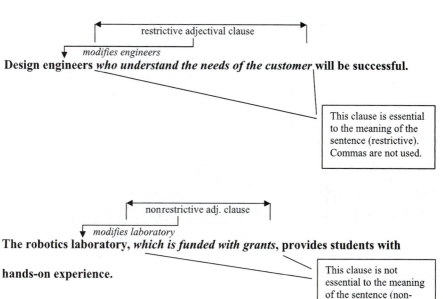

restrictive adjectival clause

modifies engineers

Design engineers *who understand the needs of the customer* **will be successful.**

This clause is essential to the meaning of the sentence (restrictive). Commas are not used.

nonrestrictive adj. clause

modifies laboratory

The robotics laboratory, *which is funded with grants,* **provides students with hands-on experience.**

This clause is not essential to the meaning of the sentence (nonrestrictive). Commas are used.

4.5 Levels of Comparison for Adjectives

Adjectives and adverbs (see Adverbs, Chapter 6), except absolute adjectives, have three comparative levels of quality or amount: *standard, comparative,* and *superlative.* For *regular adjectives,* the *standard,* also called the *positive,* is the baseline to which the others are compared. The *comparative* is either greater or less than the standard and is formed by putting *-er* at the end of the adjective, or by adding the word *more.* It is used to compare two items. The *superlative* is either the greatest or the least of the quality or amount named. It is formed by putting *-est* at the end of the adjective, or by adding the word *most.* It is used to compare three or more items. Irregular adjectives change their spelling to show the same comparative levels (see Table 4.3). If you are unsure of which form to use for any word's comparative levels, consult a good dictionary. Note: The superlative is normally used with the definite article *the* because only one superlative can exist in a given context. You could write about "*the* most important rule," but never about "*a* most important rule."

Table 4.3 Comparative levels for some regular and irregular adjectives

Regular adjectives		
Standard	Comparative	Superlative
educated	more educated	most educated
dumb	dumber	dumbest
careful	more careful	most careful
sick	sicker	sickest
awful	more awful	most awful
Irregular adjectives		
good	better	best
bad	worse	worst
much	more	most
many	more	most
some	less	least
some	more	most

05

Verbs

A *verb* is the part of speech used to denote action. We use verbs to show everything and anything that happens in time and space. Verbs always change their form to indicate <u>tense</u> (*past, present, future*). Verbs sometimes change their form to indicate <u>person</u> (*first, second, third*); <u>number</u> (*singular, plural*); and <u>mood</u> (*declarative, interrogative*, imperative, subjective*). Verbs can also be *transitive* (the verb's action requires an object) or *intransitive* (the verb's action does not require an object), and transitive verbs also have different forms for <u>voice</u> (*active* and *passive*). All of this may sound complicated, because to some extent, it is; however, verb structures in English generally make sense, which is more than one can say for some of the other structures in English. With a little practice, you should be able to deal effectively with verbs.

5.1 Definition and Function

Tense refers to the form of a verb that shows its function with respect to time. In other words, tense tells the reader whether the action occurs

5.2 Tense

---

*The *interrogative* mood is no longer included in modern presentations of English grammar. Since many other languages still differentiate the interrogative mood, however, it is being included in this book to provide a parallel structure, which might be especially useful for ESL students.

in the past, present, or future, and it also characterizes to some extent the action's flow through time.

For example, in the following sentence, the verb *to program* is in the simple past tense. The action of the verb has been completed, and the programming is finished.

The computer analyst *programmed* a new procedure.

> **Simple past tense**: action is completed.

The next example uses the *past progressive* tense. Here, the action not only occurs in the past and is complete, but during its occurrence, the verb's action showed progression. In other words, when the action occurred in the past, it represented not a "slice of time," but rather, a forward evolution of the action over a period of time.

The computer analyst *was programming* a new procedure.

> **Past progressive tense**: action progressed over a period of time but is now complete.

The following sentence provides an example of the *past perfect* tense. Here, the completed action occurs in the past <u>before</u> another event in the past, in this case, *going home.*

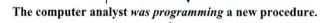

Programming occurs before *going home.*

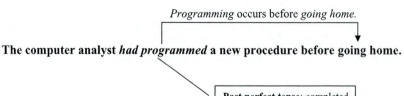

The computer analyst *had programmed* a new procedure before going home.

> **Past perfect tense**: completed action occurs before another event in the past.

Table 5.1 provides a listing of all tenses, including *simple, progressive,* and *perfect* forms, along with a description of what they do and an example of their use. Notice that the *progressive* form uses the verb *to be* and an *-ing* ending to show action in progress. The perfect tense uses a form *to have,* along with the *past participle* of the verb.

Table 5.1 Verb tenses and their functions*

Tense	Function	Example
Present simple	Shows that the action occurs now or habitually.	The Village Thumpers *sing* inspirationally.
Present simple progressive	Shows that the action is occurring progressively now.	The Village Thumpers *are singing* inspirationally at the concert.
Past simple	Shows that the action occurred in the past and is now completed.	The Village Thumpers *sang* inspirationally yesterday.
Past simple progressive	Shows that the action occurred progressively in the past and is now completed.	The Village Thumpers *were singing* inspirationally yesterday.
Future simple	Shows that the action will occur in the future but is not occurring now.	The Village Thumpers *will sing* inspirationally tomorrow.
Future simple progressive	Shows that the action will be occurring progressively in the future but is not occurring now.	The Village Thumpers *will be singing* inspirationally tomorrow.
Present perfect	Shows that the action started in the past and still continues.	The Village Thumpers *have sung* inspirationally for the past 5 hours.
Present perfect progressive	Shows that the action started occurring progressively in the past and continues to do so now.	The Village Thumpers *have been singing* inspirationally since 1999.
Past perfect	Shows that the action occurred before another event in the past and is now completed.	The Village Thumpers *had sung* inspirationally before passing out.
Past perfect progressive	Shows that the action occurred progressively before another event in the past and was completed before that event.	The Village Thumpers *had been singing* inspirationally for three hours before passing out.
Future perfect	Shows that the action will occur in the future before another event or scheduled time but is not occurring now.	The Village Thumpers *will have sung* inspirationally by the end of tomorrow's concert.
Future perfect progressive	Shows that the action will occur progressively in the future before another event or scheduled time but is not occurring now.	The Village Thumpers *will have been singing* inspirationally for several hours before sunrise.

*Note: You may encounter a few irregular exceptions to the structure provided in this table. For example, *I am going to work tomorrow* states a future action, while its construction is present progressive; and *I have just come from work* states a past action, while its construction is present perfect.

**5.3
Person
and Number**

Person is the form of the verb that denotes whether the subject of the sentence is speaking (*first person*), spoken to (*second person*), or spoken about (*third person*). *Number* is the form of a verb that denotes whether the subject of the sentence includes one (*singular*) or more than one (*plural*) person, place, thing, concept, etc. (see Table 5.2).

Table 5.2 Person and number

	Singular	Plural	Example
First person	I am I was I will be	we are we were we will be	I am certain (singular) that we are in trouble (plural).
Second person*	you are you were you will be	you are you were you will be	You are my boss (singular) while you are my staff (plural).
Third person**	he, she, or it is he, she, or it was he, she, or it will be	they are they were they will be	He is dumb (singular), but they are smart (plural).

*Exactly the same form of *you* is both singular and plural.
**Only the third person singular is gender specific.

**5.4
Irregular Verbs**

Irregular verbs are verbs that do not form their past tense by adding *-d* or *-ed*, and that may not form the past participle by adding *-d* or *-ed*.

I *live* where she *lived* and where all of her family *has lived.*

> **Regular past** ends in *-ed.*

> **Regular past participle** ends in *-ed.*

I *know* now what she *knew* yesterday and what everyone else *has known* for years.

> **Irregular past** does not end in *-ed.*

> **Irregular past participle** does not end in *-ed.*

Examples of irregular verbs are provided in Table 5.3.

Table 5.3 Examples of common irregular verbs*

Verb stem	Past tense	Past participle	Verb stem	Past tense	Past participle
arise	arose	arisen	become	became	become
begin	began	begun	bite	bit	bitten/bit
blow	blew	blown	break	broke	broken
bring	brought	brought	burst	burst	burst
catch	caught	caught	choose	chose	chosen
come	came	come	dive	dived, dove	dived
do	did	done	draw	drew	drawn
drink	drank	drunk	drive	drove	driven
eat	ate	eaten	fall	fell	fallen
find	found	found	fly	flew	flown
freeze	froze	frozen	get	got	got, gotten
give	gave	given	go	went	gone
grow	grew	grown	hide	hid	hidden
hold	held	held	keep	kept	kept
know	knew	known	lay	laid	laid
lead	led	led	leave	left	left
pay	paid	paid	prove	proved	proved, proven
ride	rode	ridden	ring	rang	rung
rise	rose	risen	run	ran	run
see	saw	seen	shake	shook	shaken
spring	sprang, sprung	sprung	swim	swam	swum
take	took	taken	tear	tore	torn
wear	wore	worn	write	wrote	written

*Consult a dictionary to determine the form of any irregular verb.

**5.5
Form
and Voice**

5.5.1 Transitive and Intransitive Forms

Verbs can take two forms in English: *transitive* and *intransitive*. Transitive verbs require an object to receive their action, while intransitive verbs do not. In a *transitive* sentence, the pattern is SUBJECT—VERB—OBJECT.

subject   verb   direct object

TRANSITIVE: **The software *controlled* the system flawlessly.**

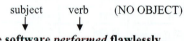

> The verb's action *controlled*
> requires the direct object *system.*

In an *intransitive* sentence, the action flows from the subject through the verb, but is not received by an object. The pattern is SUBJECT—VERB.

subject   verb   (NO OBJECT)

INTRANSITIVE: **The software *performed* flawlessly.**

> The verb's action does not
> require an object.

5.5.1.1 Linking Verbs

As mentioned in Chapter 4, *linking verbs* act like an equal sign in a sentence. They equate the subject to the part of the predicate that follows the verb. That is why linking verbs do precisely what their name indicates: they link subjects with the rest of the predicate, which, in this case, is always a *predicate noun* or *predicate adjective*, and *never an object*. Consequently, these verbs <u>are always intransitive</u> because they do not require an object.

Linking verbs basically include all forms of the verb *to be* plus all of the *sense* verbs.* Sense verbs are verbs that relate to the five senses. The following two sentences demonstrate how linking verbs are typically used.

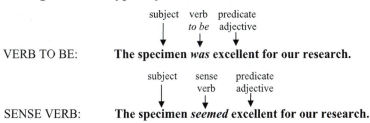

VERB TO BE: **The specimen *was* excellent for our research.**

SENSE VERB: **The specimen *seemed* excellent for our research.**

Table 5.4 provides a listing of common linking verbs.

Table 5.4 Common linking verbs

Forms of to be	Sense verbs
am	feel
are	look
is	seem
was	smell
were	sound
will be	taste

5.5.1.2 Lay *and* Lie

Many times, confusion over which verb to use actually involves confusion over transitive and intransitive forms. For example, the transitive verb *to lay* is often confused with the intransitive verb *to lie*. When we *lay* something, we are putting or placing <u>it</u> somewhere. The verb

---

*Some sense verbs may not always be used as linking verbs. For example, in the sentence *John feels the cold*, the verb *feels* is a sense verb taking the direct object *cold*. Here, *feels* is not a linking verb.

requires a direct object, in this case, *it*. When we
lie, we are assuming a flat or reclining position.
No object is necessary.

TRANSITIVE: **I will *lay* my head on the pillow when finished.**

> The verb's action requires
> an object (*head*).

INTRANSITIVE: **I will *lie* down when finished.**

> The verb's action does not
> require an object.

BOTH: **I *lay* bricks when working and just *lie* around when I am home.**

> *Transitive* form
> requires an object.

> *Intransitive* form does
> not need an object.

5.5.2 Active and Passive Voice

Voice is simply how the verb relates to its subject. If the doer of the action comes before the
verb, then the sentence is in the *active voice*. If
the receiver of the verb's action comes before the
verb, then the sentence is in the *passive voice*.
Transitive sentences can be either *active* or *passive*. Intransitive sentences can only be active
since no receiver of the action exists to precede
the verb.

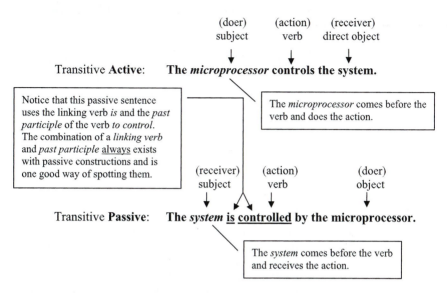

Passive sentences effectively reverse the functions of the subject and direct object. Subjects, which normally do the action, now receive the action; and objects, which normally receive the action, if present in the sentence, now do the action. Passive sentences also tend to use wordy and weak constructions, while active sentences are more direct and efficient. Remember: *active* sentences always have the "doer" of the verb's action before the verb and, if transitive, always have the receiver of the verb's action after the verb. *Passive* sentences, which are always transitive, have the subject effectively functioning as a direct object before the verb, and may or may not have the doer of the action after the verb.

5.5.2.1 Passive Voice in Technical and Scientific Writing

For most writing, active voice provides a more direct, energetic, and efficient style; however, especially in technical and scientific writing, passive voice has its place because it can actually be more efficient and effective in certain situations. Also—right or wrong, good or bad—technical and scientific audiences *expect* passive voice in technical writing and some even equate it with being professional. There are a couple of reasons for this:

- **Implied objectivity.** *Technical and scientific writing is supposed to be objective.* Passive voice, by removing the doer, seems more objective. "I collected the data" does not sound as impartial as "The data were collected." Right or wrong, not being perceived as objective by a scientific audience can compromise the value of your work. A good rule is to always know what is expected (including what writing style is expected) <u>before</u> doing any kind of technical writing.
- **Irrelevance of the subject/importance of the object**. In many cases, in technical and scientific writing, the doer of the action is unimportant or irrelevant. For example, if you were describing the steps of a laboratory procedure for making a compound, you might say, "this is added to that," then, "the mixture is heated to a certain temperature for so long," etc. The doer in this situation (maybe a lab technician performing the work) is not important, but the receiver of the action (the procedure being performed) is. So in situations where *what is done* is the focus, <u>not</u> who is doing it, passive voice might actually make good sense.

5.5.2.2 Making the Passive Voice Active

To make a passive sentence active: (1) delete the linking verb, (2) put the verb's complement (which will now include the direct object) after the verb, and (3) move the doer of the action (which will now act as the subject) to a position before the verb. Note: if you do not know the doer, you will need to create one.

How to activate a passive sentence

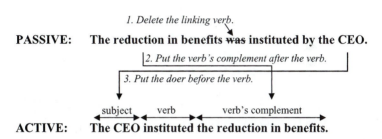

1. Delete the linking verb.

PASSIVE: The reduction in benefits ~~was~~ instituted by the CEO.

2. Put the verb's complement after the verb.

3. Put the doer before the verb.

subject   verb   verb's complement

ACTIVE: The CEO instituted the reduction in benefits.

What do we mean by the *mood* of a verb? Happy verbs? Sad verbs? Apprehensive verbs? If mood were only that easy to define! The *mood* of a verb is normally defined either as the attitude of the speaker or writer, or the speaker or writer's conceptual process for using an action or state. Unfortunately, such definitions do not help anyone understand what mood is or how it should be used.

It might be more useful to describe what *mood* does rather than ponder what *mood* is. The four moods we will discuss are *declarative, interrogative, imperative,* and *subjunctive*. As mentioned earlier, many scholars today no longer consider *interrogative* to be a separate mood. However, it is included here to clarify, for native speakers of other languages, the relationship of verb mood in English to similar structures in their native languages.

**5.6
Mood**

The mood of a verb tells the reader whether the sentence expresses a fact or makes a statement (*declarative*), asks a question (*interrogative*), gives a command (*imperative*), or expresses something that is conditional or contrary to fact (*subjunctive*).

5.6.1 Declarative Mood

The *declarative* mood (also called the *indicative* mood) expresses a fact or makes a statement about something or someone.

The following sentence is declarative because it makes factual statements. Declarative verbs follow standard rules, and the sentence usually ends with a period.

Gene sequencing *requires* a high-end processor.

> **Declarative:**
> Provides a statement about gene sequencing.

5.6.2 Interrogative Mood

The following sentence is *interrogative* because it asks a question. Verbs of interrogative sentences follow standard rules, and the sentence ends with a <u>question mark</u>.

***Does* gene sequencing require a high-end processor?**

> **Interrogative:**
> Asks a question about gene sequencing and processors.

5.6.3 Imperative Mood

The following sentence is in the *imperative* mood because it gives a command. Verbs of imperative sentences often lead off a sentence that ends with a period. Because the subject is the person or persons being addressed, the pronoun *you* is understood and is usually not included in the sentence.

> "You" is understood to be the subject.

> Imperative: gives a command to obtain a high-end processor.

YOU Obtain **a high-end processor to do gene sequencing.**

5.6.4 Subjunctive Mood

Over the years, the subjunctive mood has been shrouded in mystery. Only the shy, introverted English teacher lurking in the shadows really knew. But the subjunctive is no mystery—it is not that difficult to understand. In fact, virtually any engineering or science student who has written a computer program is already familiar with the basic concept behind the subjunctive mood. Looping code such as IF-THEN, WHILE-WEND, or DO-UNTIL is all based on expressing a conditional statement. That is also what subjunctive verbs do. Specifically, they express statements contingent on three general and somewhat overlapping conditions. Verbs are in the subjunctive mood when they:

- Express a condition based on a wish, desire, or hope.
- Express a condition that is contrary to fact, hypothetical, or highly unlikely.
- Express a condition that is asking, requesting, or demanding something.

Subjunctive verbs use the base form of the verb (the form of a verb listed first in the dictionary), or in the case of the verb *to be,* either *be* or *were.* The subjunctive form of the verb may appear distinctive and unusual compared with verbs in other moods. For example, the subjunctive frequently pairs a singular subject with the past plural form of *to be.*

The sentence below states a condition of wish, desire, or hope. Specifically, the subordinate clause is contingent on a wish expressed in the main clause and is therefore subjunctive.

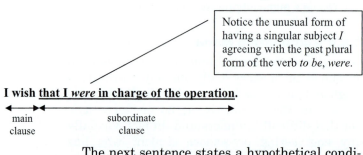

Notice the unusual form of having a singular subject *I* agreeing with the past plural form of the verb *to be, were.*

I wish that I *were* in charge of the operation.

main clause   subordinate clause

The next sentence states a hypothetical condition that is contrary to fact and may even represent wishful thinking. The subordinate clause beginning with *if* forms this hypothetical condition and is therefore subjunctive.

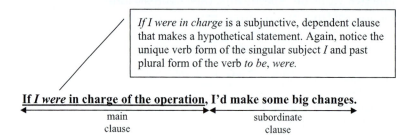

If I were in charge is a subjunctive, dependent clause that makes a hypothetical statement. Again, notice the unique verb form of the singular subject *I* and past plural form of the verb *to be, were.*

If *I were* in charge of the operation, I'd make some big changes.

main clause   subordinate clause

The final sentence provides an example of the subjunctive mood formed by the expression of a question, request, or demand. The subordinate clause is conditioned on a request expressed in the main clause and is therefore subjunctive.

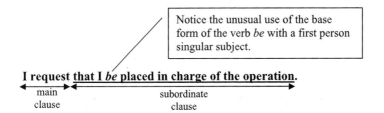

Notice the unusual use of the base form of the verb *be* with a first person singular subject.

I request <u>that I *be* placed in charge of the operation</u>.

main clause   subordinate clause

06
00110

Adverbs

Adverbs are the words we generally use to modify, describe, or limit a verb. They tell us *how, when, where, which one, how many,* or *what quality*. Besides verbs, we can also use adverbs to modify adjectives, other adverbs, phrases, complete clauses, and verbals (verbs used as nouns, adjectives, or adverbs—i.e., gerunds, participles, and infinitives). Adverbs often end in *-ly,* but not always.

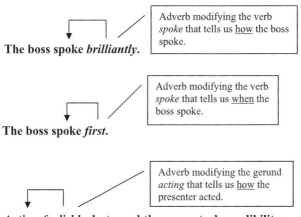

Adverb modifying the verb *spoke* that tells us <u>how</u> the boss spoke.

The boss spoke *brilliantly.*

Adverb modifying the verb *spoke* that tells us <u>when</u> the boss spoke.

The boss spoke *first.*

Adverb modifying the gerund *acting* that tells us <u>how</u> the presenter acted.

Acting *foolishly* destroyed the presenter's credibility.

6.1 Adverbials

An adverbial can be any word, phrase, or clause that is being used as an adverb.

6.1.1 Adverbial Word

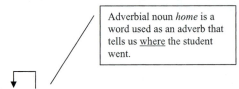

Adverbial noun *home* is a word used as an adverb that tells us <u>where</u> the student went.

The exhausted engineering student went *home.*

6.1.2 Adverbial Phrase

An *adverbial phrase* is a group of words that lacks either a subject or predicate and that acts as an adverb in a sentence. Prepositional and infinitive phrases frequently function as adverbial phrases.

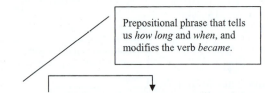

Prepositional phrase that tells us *how long* and *when*, and modifies the verb *became.*

Within a few years, the Internet <u>became</u> a commercial network.

6.1.3 Adverbial Infinitive Phrase

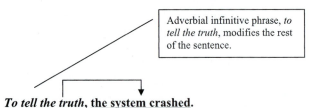

Adverbial infinitive phrase, *to tell the truth*, modifies the rest of the sentence.

To tell the truth, <u>the system crashed.</u>

6.1.4 Adverbial Split-Infinitive Phrase

An infinitive consists of *to* and a verb stem. Splitting the infinitive by putting a word between *to* and the verb stem was once frowned upon; however, modern usage allows split infinitives either when they read or sound better, or clarify meaning.

> Adverbial infinitive phrase, *to tell the truth,* is split by the adverb *really.* This split infinitive is acceptable here because changing the sentence to read, *"To tell really the truth,"* or *"Really, to tell the truth,"* changes the meaning and readability of the sentence.

To really tell the truth, **I crashed the system.**

One parenthetical clarification is needed here regarding this phrase. *Truth* is normally considered an absolute quality (see 4.2.6, 6.1.5); however, this construction is idiomatic (see 7.2.2) and reflects not on the degree of truth being told, but rather, on the *degree of truthfulness or accuracy* of the teller. As we all know, *truthfulness* and *accuracy* are not absolutes.

6.1.5 Absolute Phrase

An *absolute phrase* is an adverbial phrase consisting of a noun and participle that modifies the entire sentence, but it is not connected to the sentence with either a conjunction or relative pronoun. Also called a *nominative absolute*, absolute phrases are always parenthetical to the sentence they modify.

> This absolute phrase contains the noun *computer* and the participle *flooded.* The phrase is parenthetical to the main clause that follows.

The computer having been flooded with bogus requests, **the processing of our data was delayed for hours.**

6.1.6 Adverbial Clause

Also called an *adverb clause*, it is a subordinate clause used as an adverb in a sentence.

This subordinate clause tells us *why* and *when* the Internet evolved rapidly and thus functions as an adverb in this sentence.

The Internet evolved rapidly *when high-speed service became commonplace.*

**6.2
Levels
of Comparison
for Adverbs**

Like adjectives (see 4.5), adverbs have three comparative levels of quality or amount: *standard, comparative,* and *superlative.* As with adjectives, the comparative is either greater or less than the standard and is formed by putting *-er* at the end of the adverb or by adding the word *less* or *more* before the adverb. The comparative is used when comparing two items. The superlative is either the greatest or the least and is formed by putting *-est* at the end of the adverb or by adding the word *least* or *most* before the adverb. The superlative is used when comparing three or more items. Irregular adverbs change their spelling to show the same comparative levels (see Table 6.1).

Table 6.1 Examples of regular and irregular adverbs*

Regular adverbs		
Standard	**Comparative**	**Superlative**
soon	sooner	soonest
quickly	more/less quickly	most/least quickly
carefully	more/less carefully	most/least carefully
Irregular adverbs		
well	better	best
badly, poorly	worse	worst

*When in doubt as to the formation of comparative levels for a particular adverb, always consult a good dictionary.

Compound constructions normally involve an adverb paired with, and modifying, an another adverb.

6.3 Compound and Absolute Adverbs

The adverb *very* modifies the adverb *foolishly* in this compound structure. Together they tell us how the boss spoke.

The boss spoke *very foolishly*.

Do not confuse compound adverbs with compound adjectives, which involve two adjectives each of which modifies the noun that follows (see 4.2.4).

6.3.1 Adverbs with Absolute Adjectives

Certain adverbs can be used to modify absolute adjectives in compound constructions in a way that modifies but <u>does not change</u> the adjective's absolute meaning. For example, you cannot be a *very dead* person or an *extremely dead* person, since the state of being dead is absolute. But you can be an *almost dead* person or *nearly dead* person, meaning the person is approaching the absolute state of *dead*, but has not quite reached it (see 4.2.6).

The placement of adverbs in a sentence can affect the meaning of the sentence. In the following example, the location of the adverbial phrase, *in basements late at night,* is too far removed from the subject *hackers,* which it is supposed to modify. The result: the sentence seems to say that hackers only compromise systems located in basements and only late at night. It should say

6.4 Placement of Adverbs

that the hackers, not the systems, are in basements late at night.

Hackers can compromise systems anywhere *in basements late at night.*

This sentence is another example of dangling modifiers, which were discussed in an earlier chapter in terms of misplaced adjectival phrases (see 4.1.3). The physical placement of an adverb in a sentence can change what it modifies, and in the process, it can change the entire meaning of the sentence.

To fix the problem, simply move the adverbial prepositional phrase nearer to *hackers.*

Hackers, *in basements late at night,* can compromise systems anywhere.

6.4.1 Adverb Categories and Placement

Adverbs tend to fall into four general categories:

- Adverbs of *frequency* that describe how often.
- Adverbs of *degree* that describe how much.
- Adverbs of *manner* that describe way, mode, method, or style.
- Adverbs of *limitation* that restrict or restrain.

Table 6.2 provides general guidelines for placement of adverbs organized by these general categories; however, adverbs can be placed properly almost anywhere in a sentence. Consequently, adverbs do not always follow the placement scheme provided in this table, but they usually do.

Table 6.2 Adverb categories and placement

Adverb category/examples	Placement	Example
Adverbs of frequency *always, never, often, rarely,* *seldom, sometimes, usually*	First of the sentence	*"Usually,* computers increase productivity."
Adverbs of degree *absolutely, almost, certainly,* *completely, definitely,* *especially, hardly, only*	Before the word that is modified	"Computers have *certainly* improved productivity."
Adverbs of manner *badly, beautifully, best,* *openly, lightly, well*	After the verb	"Computers work *best* with purely computational tasks."
Adverbs of limitation *only, nearly, simply,* *scarcely, merely*	Before the word being modified	*"Only* computers can save the world."

6.5 Transitional Phrases and Adverbial Conjunctions

Adverbs are often used to connect main clauses. Since they function as coordinating conjunctions when used in this way, they are called *adverbial conjunctions* or *conjunctive adverbs*. Additionally, certain adverbial phrases, called *transitional phrases*, can also be used to join main clauses.

6.5.1 Punctuating Conjunctive Adverbs and Transitional Phrases

Adverbial conjunctions, such as *however* and *therefore*, connect main clauses and are preceded with a semicolon and followed by a comma. Transitional phrases, such as *that is* or *what is more,* serve the same function and are punctuated the same way. Table 6.3 lists common conjunctive adverbs, while Table 6.4 provides examples of transitional phrases.

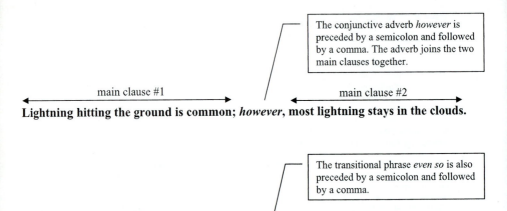

The conjunctive adverb *however* is preceded by a semicolon and followed by a comma. The adverb joins the two main clauses together.

main clause #1 ← → main clause #2 ← →

Lightning hitting the ground is common; *however*, most lightning stays in the clouds.

The transitional phrase *even so* is also preceded by a semicolon and followed by a comma.

main clause #1 ← → main clause #2 ← →

Lightning hitting the ground is common; *even so*, most lightning stays in the clouds.

Table 6.3 Common conjunctive adverbs

accordingly	hence	moreover
also	henceforth	nevertheless
anyhow	however	next
anyway	incidentally	otherwise
besides	indeed	still
consequently	instead	then
finally	likewise	therefore
furthermore	meanwhile	thus

Table 6.4 Examples of transitional phrases

after all	in some way	on the other hand
at the same time	in spite of	that is
even so	in the same way	that notwithstanding
for example	of course	to illustrate
for instance	on the contrary	what is more

CHAPTER SEVEN

Prepositions

A *preposition* is the part of speech we use to show the relationship of a noun or noun substitute with another word in a sentence. A preposition is normally used with a noun or pronoun in a word group called a *prepositional phrase*. For example:

> Prepositional phrases that modify the system's *functioning*

The system is functioning *in* the mode *for which* it was designed.

The first preposition *in* has the noun *mode* as its object. The second preposition *for* has the pronoun *which* as its object. These phrases tell us something about the relationship between the noun *system* and the verb *designed*. Some of the most common prepositions are provided in Table 7.1.

Table 7.1 Common prepositions

about	beside	inside of	round
above	between	in spite of	since
according to	beyond	into	through
across	by	like	throughout
after	concerning	near	till
against	despite	next to	to
along	down	of	toward
along with	due to	off	under
among	during	on	underneath
around	except	onto	unlike
as	except for	on top of	until
at	excepting	out	up
because of	for	out of	upon
before	from	outside	up to
behind	in	over	with
below	in addition to	past	within
beneath	inside	regarding	without

7.1 Uses of Prepositional Phrases

As mentioned, a preposition combines with a noun or pronoun to form a *prepositional phrase*. When included as part of a sentence, prepositional phrases are most often used as adjectives and adverbs.

7.1.1 Use as an Adverb

Here, the prepositional phrase acts as an adverb because it functions to tell us *where* the testing occurs.

Testing the engine *in* a propulsion laboratory is one effective approach.

Adverbs generally modify verbs, verbals, adjectives, and other adverbs to tell us where, when, or how something is happening. In the sentence above, the prepositional phrase, *in the labora-*

tory, tells us where the testing takes place. Consequently, the prepositional phrase functions as an adverb.

7.1.2 Use as an Adjective

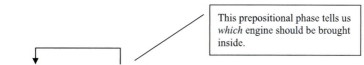

This prepositional phase tells us *which* engine should be brought inside.

The engine _on the flatbed truck_ really ought to be brought inside.

Although, as with the earlier example, the prepositional phrase specifies *where*, its primary function in this sentence is to tell us *which* engine—the <u>one</u> on the flatbed truck. Because it modifies the noun *engine,* the phrase primarily functions as an adjective.

For ESL students, prepositions bring both good and bad news. The good news is that they never change their form. So much for the good news! The bad news is that the meanings of prepositions often change significantly and illogically when they are used with certain verb constructions—especially the two-word verb forms commonly found in spoken English and informal writing. A *two-word verb* is a combination of a verb and a preposition that functions together as a verb. Prepositions used like this are often called *particles*. Changing the preposition usually changes the entire meaning of the verb, as well as the sentence in which it functions.

7.2 Prepositions as a Part of Two-Word Verbs

Consider the meanings of the following two sentences and notice how changing the preposition changes the entire sense of what the sentence says.

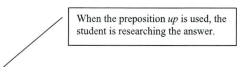

When the preposition *up* is used, the student is researching the answer.

The student is *looking up* the answer to the problem.

When the preposition *over* is used, the student has already provided an answer and is now reviewing it.

The student is *looking over* the answer to the problem.

7.2.1 Splitting Two-Word Verbs

In the following example, notice the prepositions *down* and *up*. This sentence contains a pair of two-word verbs (*cut down* and *cut up*). Each consists of a verb (*cut*) and a preposition (*down* or *up*). Notice also that the two-word verbs have been split with the noun *tree* (direct object) in the first clause, and with the pronoun *it* (direct object) in the second clause.

Tree and *it* split the two-word verb.

The man *cut* the tree *down*, and then he *cut* it *up*.

Most two-word verbs that are transitive (those that take an object) can be split if doing so sounds better and improves readability and clarity. Table 7.2 provides a list of some common verb-preposition combinations that can be split. Table 7.3 lists common verb-preposition combinations that cannot be split.

Table 7.2 Two-word verbs that can be split

bring up	give away	point out	throw out
call off	hand in	print out	try on
call up	hand out	put away	try out
drop off	help out	put back	turn down
fill out	leave out	put off	turn on
fill up	look over	take out	turn up
gets under	make up	take over	wrap up

Table 7.3 Two-word verbs that cannot be split

catch on	go over	play around	stay away
come across	grow up	run into	stay up
get along	keep on	run out of	take care of
give in	look into	speak up	turn up at

Perhaps a more important consideration in this last example is how absurd this sentence might seem to someone interpreting the language literally: *down* means toward a lower position, while *up* means toward a higher position. Cutting the tree down makes sense—when you cut the tree, it falls down. But, what about cutting the tree up? Does that mean the tree just sort of jumps up off the ground and stands tall again when you cut it in the other direction? Clearly, *cutting up a tree* after it has been cut down makes no literal sense at all. Since two-word verb combinations like *cut up* can have several meanings separate from the literal definitions of the words themselves, they are said to be *idioms*. For example, when the author was a young child, he was routinely sent to his room for *cutting up* (misbehaving) in front of guests.

7.2.2 Idiomatic Usage

Idioms are accepted phrases, constructions, or expressions that are contrary to the usual pattern of English and that have a meaning different from the literal meaning. While native speakers of English usually have little or no trouble with idioms, nonnative speakers do. This is because no consistent set of rules exists for determining if something is an idiomatic expression, and if so, what it actually means.

There is an old adage that an author who tries to explain idioms is on a fool's errand. That being said, it is important to at least note that prepositions are frequently used as key parts of *idiomatic expressions*. These expressions are formed from words that mean something different from what the words by themselves would indicate. Idioms also may not follow normal rules of grammar.

For example, consider the two-word verb *got under* in the contexts of the following two examples. In the first example, the professor's long, boring lecture is <u>irritating</u>. The meaning of the words *got under* is totally different from what the words would normally indicate; consequently, this construction is considered to be idiomatic.

> The prepositional phrase formed by the two-word verb is idiomatic.

The professor *got under* my skin with his long, boring lecture.

In the second example, the same two-word verb is used literally.

> Here the same prepositional phrase is literal. The physician took a scalpel and actually did get under the skin.

The surgeon's scalpel *got under* my skin when she removed a cyst.

Notice that the only difference between the two examples is the context: the first is about a boring professor, while the second involves a surgical procedure. An ESL student would have no problem with the second example, but the first example could be difficult to understand. For a native speaker, however, the prepositional phrase in the first sentence would make perfect sense. When one grows up immersed in a language, idioms come naturally; however, for non-native speakers, idioms can be very difficult.

No easy answers exist for ESL students where idioms are concerned, except to watch out for two-word verbs in prepositional phrases that do not seem to make literal sense. Of course, many idioms also make literal sense, but their literal sense is wrong. For example, what if a politician speaking to 500 people at a campaign rally asked the audience to, "Lend me your ears"? Literally, he or she is asking the audience to hand over 1,000 ears. Probably the resulting collection of body parts would not enhance the campaign. The idiomatic meaning, which is the real meaning, is for the audience to listen to what he or she has to say.

7.2.3 Unidiomatic Consequences

Sometimes you just cannot win, especially where idioms are concerned. Even though idioms are frequently illogical and not grammatically correct, if you are logical and grammatically correct when you should be idiomatic, then bad things can happen. Look at the following three sentences, each of which combines the verb *agree* with a different preposition in a two-word verb construction. Notice how the choice of preposition effectively controls the idiomatic meaning of the sentence.

We approve of these conditions.

We agree <u>with</u> the conditions of the contract.

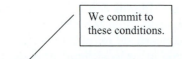

We commit to these conditions.

We agree <u>to</u> the conditions of the contract.

We understand these conditions.

We agree <u>on</u> the conditions of the contract.

Because a preposition's use in two-word verbs is frequently idiomatic, ESL students often have problems saying what they intend to say. The best approach is to listen and deduce intended meanings by context. Also check out the many excellent Websites devoted to idioms in English, such as: http://home.t-online.de/home/toni.goeller/idioms/.

08
01000

Conjunctions

Conjunctions are the words we use to connect words, phrases, clauses, and sentences. Conjunctions are normally divided into four general categories: *coordinating conjunctions, correlative conjunctions, subordinating conjunctions*, and *adverbial conjunctions*. Conjunctions are punctuated differently depending on how they are used.

8.1 Coordinating Conjunctions

Coordinating conjunctions are used to connect words, phrases, or clauses of equal rank or importance. See Table 8.1.

Table 8.1 Coordinating conjunctions

and	or	for
but	nor	yet

8.1.1 Words

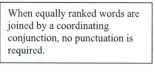

When equally ranked words are joined by a coordinating conjunction, no punctuation is required.

The road was paved with a mixture of <u>asphalt</u> *and* <u>gravel</u>.

8.1.2 Phrases

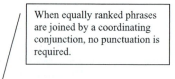

When equally ranked phrases are joined by a coordinating conjunction, no punctuation is required.

The computer programmer ran **<u>down the hill</u>** *and* **<u>into the woods</u>**.

8.1.3 Clauses

When equally ranked main or independent clauses are joined by a coordinating conjunction, a *comma* precedes the conjunction.

<u>A fault developed in the power stage,</u> *but* <u>the amplifier continued to function.</u>

**8.2
Correlative
Conjunctions**

Correlative conjunctions are pairs of conjunctions used to connect parallel structures. Often, correlative conjunctions are pairs of coordinating conjunctions, but not always. See Table 8.2.

Table 8.2 Correlative conjunctions

both ... and	neither ... nor	whether ... or
either ... or	not only ... but also	

When *correlative conjunctions* join words or phrases, no mark of punctuation is required.

The workstation's problems included *both* <u>excessive glare</u> *and* <u>inadequate chair height</u>.

When a *correlative conjunction* joins independent clauses, the conjunction joining the clauses should be preceded by a comma.

Either <u>you will do the work,</u> *or* <u>I will replace you with a machine</u>.

Subordinating conjunctions are used to join two clauses together, and in the process, subordinate one of the clauses to the other. See Table 8.3.

8.3 Subordinating Conjunctions

Table 8.3 Subordinating conjunctions

as if	if	when
although	since	where
because	such as	while

Because joins the two clauses and also subordinates the clause that follows it. Unless the readability of the sentence requires it, no punctuation is needed.

<u>The amplifier circuit shut down</u> *because* <u>a fault developed in the power stage</u>.

8.4 Adverbial Conjunctions

Adverbial conjunctions, also called *conjunctive adverbs*, are adverbs that are being used to join main clauses. Additionally, certain phrases, called *transitional phrases*, can function as adverbial conjunctions (see also Section 6.5, Table 6.3, and Table 6.4).

8.4.1 Conjunctive Adverb

> The conjunctive adverb *however* is preceded with a semicolon and followed by a comma.

The wing box failed due to corrosion and fatigue; *however*, the pilot ejected safely.

8.4.2 Transitional Phrase

> The transitional phrase *what is more* functions like a conjunctive adverb. It is also punctuated like one with a semicolon and a comma.

The wing box failed due to corrosion and fatigue; *what is more*, the plane had just undergone its periodic maintenance cycle.

09
01001

Interjections

Interjections are words or groups of words we use either as an exclamation or to show surprise, emotion, or impact. We normally think of words such as *ouch*, *whew*, and *wow* as interjections. However, almost any word or group of words can be used as an interjection. Whether a word or group of words is, in fact, an interjection or some other part of speech depends solely on its function in the sentence. If the word acts in the sentence only to exclaim, then it is an interjection. Interjections are normally punctuated with an exclamation point or a comma. *In technical writing, you will rarely, if ever, use an interjection.* In other types of writing where interjections may be appropriate, be careful not to overuse exclamation points as they can lose their effect very quickly. In fact, a good rule is to use them only when they are needed to convey the writer's intent.

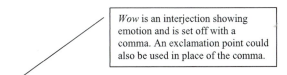

Wow is an interjection showing emotion and is set off with a comma. An exclamation point could also be used in place of the comma.

Wow, I love your new car.

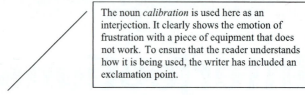

The noun *calibration* is used here as an interjection. It clearly shows the emotion of frustration with a piece of equipment that does not work. To ensure that the reader understands how it is being used, the writer has included an exclamation point.

Calibration! **We cannot even apply power to this piece of junk**.

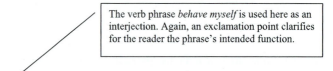

The verb phrase *behave myself* is used here as an interjection. Again, an exclamation point clarifies for the reader the phrase's intended function.

Behave myself! **You have got to be kidding**.

10
01010

Punctuation

Punctuation is the use of symbols, marks, and signs to separate or connect certain parts of a sentence and to clarify meaning. In English, punctuation marks include the following: *apostrophe, brackets, colon, comma, ellipsis, dash, hyphen, exclamation point, parentheses, period, question mark, quotation marks, semicolon,* and *slash.* Table 10.1 provides a quick reference guide to these punctuation marks, while the rest of the chapter, arranged alphabetically by punctuation mark, provides a more detailed discussion for each mark.

Table 10.1 Quick guide to punctuation

	Mark	Function	Example
Introducers	**colon** *Go to 10.3*	Introduces **with emphasis** a word, phrase, clause, or series that follows.	*Please do the following: get up, get going, and get a job.*
	dash *Go to 10.5*	Introduces **with emphasis** a parenthetical idea or new thought within a sentence.	*If this invention works— and the lab thinks it will—we will be rich.*
Separators & connectors	**comma** *Go to 10.4*	Separates items in a series.	*Soldering involves cleaning, heating, flowing, and cooling.*
		Separates nonrestrictive phrases and clauses.	*Because we care, we are donating this equipment.*
		Separates phrases and clauses with meaning that is different from what comes before.	*Do not install the new patch, not unless you're ordered to.*
		Separates main clauses with a conjunction.	*He secured the blockhouse, and she started the countdown.*
	semicolon *Go to 10.13*	Separates items in a series when any one item contains a comma.	*Please get up; get going, if you can; and get a job.*
		Separates two, closely related main clauses.	*She smiled; I frowned.*
	slash *Go to 10.14*	Separates options.	*The process resulted in a true/false output.*
	hyphen *Go to 10.8*	Divides words at the end of the line.	*Please meet me at the al-ternate location.*
		Connects certain written whole numbers and fractions.	*Forty-five percent is less than one-half interest.*
Containers	**quotation marks** *Go to 10.12*	Contains direct quotations.	*He said, "You have the job."*
		Contains minor titles such as chapters, articles, episodes.	*I think the best Seinfeld episode was "The Soup Nazi."*

	Mark	Function	Example
Containers *cont.*	**parentheses** *Go to 10.9*	Contains parenthetical material incidental to the main thought.	*We were relieved when the ion chamber worked (it never had before).*
	brackets *Go to 10.2*	Contains explanations or editorial corrections.	*Scientists tend to … [become] opinionated.*
		Contains a second level of parentheses within a level of parentheses.	*The operating system (the requested OS upgrade [Mac OS-X]) worked well.*
Terminators	**period** *Go to 10.10*	Terminates a sentence that is neither a question nor an exclamation.	*Voice Over Internet Protocol is the new telephone paradigm.*
		Used with most abbreviations.	*He has a Ph.D., experience, and solid references.*
	question mark *Go to 10.11*	Terminates a sentence that asks a question.	*Can you fund my project?*
	exclamation point *Go to 10.7*	Terminates a sentence that makes an exclamation.	*I will not take a pay cut!*
		Punctuates an interjection.	*No! I will not take a pay cut.*
Indicators	**apostrophe** *Go to 10.1*	Indicates possession.	*Phyllis' artwork*
		Indicates an omission in words or numbers.	*His career peaked in '68.*
		Pluralizes letters, abbreviations, symbols, and numerals.	*I need more 0's before the decimal point in my bank balance.*
	ellipsis *Go to 10.6*	Indicates omitted material.	*I need more 0's … in my bank balance.*

<table>
<tr><td>

10.1
Apostrophe

</td><td>

Use an apostrophe to indicate possession, omissions, and contractions, and for pluralizing letters, abbreviations, symbols, and numerals. The following subsections will provide flowcharts and other aids to help you properly use the apostrophe for its many purposes.

</td></tr>
</table>

10.1.1 Possession

To make a noun show ownership or possession, add an apostrophe or apostrophe + *s* as required by the particular construction (see 2.2). Figure 10.1 is a flowchart that shows how to make various noun constructions possessive.

10.1.2 Omissions and Contractions

Use apostrophes to substitute for omitted parts of words or numbers, especially when forming contractions.

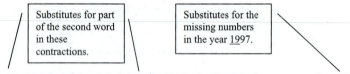

| Substitutes for part of the second word in these contractions. | Substitutes for the missing numbers in the year 1997. |

She'll attend although I didn't invite her. She has been away since '97.

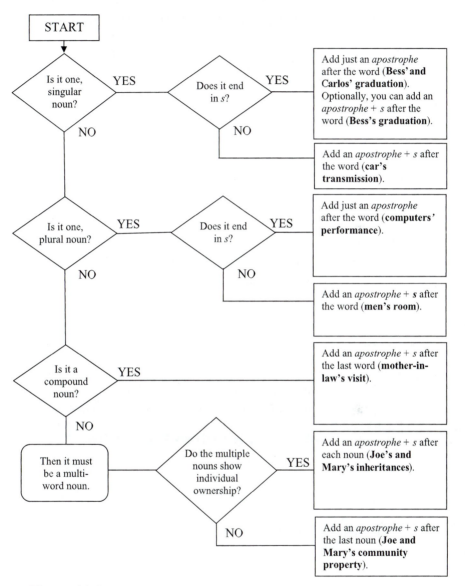

Figure 10.1
Flowchart for making nouns possessive.

10.1.3 Pluralizing Letters, Abbreviations, Symbols, and Numerals

Use apostrophes when you want to make numbers and lowercase letters plural. You can also use apostrophes to make upper case letters plural, but normally you would only do that to avoid confusing them with something else.

Apostrophes are used when pluralizing numerals.

Digital systems use 1's and 0's to encode information.

Pluralizing lowercase letters always requires an apostrophe. The apostrophe was added to the upper case A to prevent confusing it with the word, *as.*

The code sequence, *bbAAA*, contains 2 b's and 3 A's.

**10.2
Brackets**

You can use *brackets* for two purposes: (1) to set off explanations or editorial corrections within quoted material; and (2) to indicate a set of parentheses within an existing set of parentheses.

10.2.1 Setting Off an Explanation

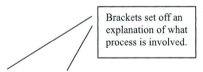

Brackets set off an explanation of what process is involved.

She said, "End that process [merge sort] if the CPU cannot keep up."

10.2.2 Setting Off a Second Level of Parentheses

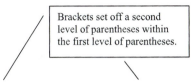

Brackets set off a second level of parentheses within the first level of parentheses.

The new boss (Joe Smith [Allen Jones' replacement]) wants to meet at 4 p.m.

You can use *colons* as a formal introduction to additional material with the intention of directing your reader's attention to that material. In this regard, colons are normally used as a stronger mark of punctuation to focus the reader's attention on direct quotations, explanations or summaries, a series of items, or appositives. *Colons* are also used in time references, between titles and subtitles, and in source documentation.

10.3 Colon

10.3.1 Direct Quotation

A colon sets off the direct quotation in a way that directs the reader's attention to it.

The lecturer said: "Synthesis is the process of creating useful, complicated things from simple, uncomplicated things."

10.3.2 Explanation

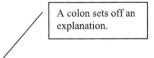

A colon sets off an explanation.

Synthesis is the basis of digital sound: it allows us to create complicated sounds from more basic sounds.

10.3.3 Series of Items

> A colon introduces items in a series.

Digital image editing is most commonly done for the following reasons: image enhancement, special effects, animation, and compositing.

10.3.4 Appositives

> A colon sets off an appositive of the word *tomography*.

Digital medical imaging often relies on tomography: a series of cross-sectional slices of the area being imaged.

10.3.5 Time References

> A colon divides hours from minutes in a time reference.

Meet me at the launch pad at 10:30 a.m. tomorrow.

10.3.6 Titles and Subtitles

> A colon separates the title of a book from the subtitle.

Stephen Hawking's book *A Brief History of Time: From the Big Bang to Black Holes* provides an advanced theory of astrophysics.

10.3.7 Source Documentation

> A colon is often used in formal source documentation.

Einstein, Albert. *Ideas and Opinions*. New York: Crown Publishers, Inc., 1982.

Use *commas* to indicate where the reader should pause briefly or change inflection when reading a sentence. Commas normally precede coordinating conjunctions that join main clauses, follow certain introductory elements, separate items in a series, and set off nonrestrictive or parenthetical elements in a sentence.

**10.4
Comma**

10.4.1 Preceding Coordinate Conjunctions

> A comma precedes the coordinate conjunction separating two main clauses.

He turned on the weather radar, and she saw the tornado vortex signature.

10.4.2 Following Certain Introductory Elements

> A comma follows an introductory adverbial phrase.

When he turned on the weather radar, she saw the tornado vortex signature.

> A comma follows a long, introductory, prepositional phrase.

In a control room full of weather displays, she saw the tornado vortex signature.

> A comma is not required after a short, introductory prepositional phrase, but can be included for clarity.

At her workstation she saw the tornado vortex signature.

> A comma follows an introductory transitional expression.

He feared the storm. In fact, he cowered in the shelter for hours.

10.4.3 Separating Items in a Series

> Commas separate items in a series when none of the items contains a comma. If one item does, then semicolons are used to separate all the items. (See 10.11, *semicolons*.)

The cowering scientist was dirty, sweaty, stinky, and nervous.

10.4.4 Setting Off Nonrestrictive Elements

> Commas are used to set off this nonrestrictive clause. The clause is nonrestrictive because it is not essential to the meaning of the sentence.

Proper grammar, <u>as my teacher used to say</u>, is a prerequisite for success in life.

You can double check your use of commas in a sentence with the flowchart provided in Figure 10.2.

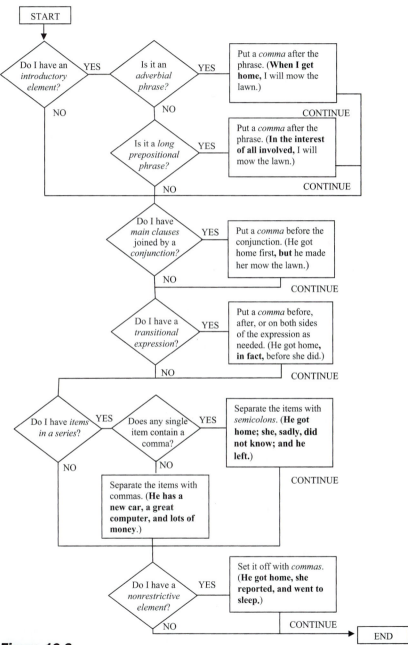

Figure 10.2
Flowchart for using commas.

10.5
Dash

Use a *dash* when you want to indicate an interruption or change in dialogue or thought, to emphasize a parenthetical element, when connecting an introductory series of words with an explanation for what the words mean, or when needed for clarity or emphasis. To produce a dash, you can use the old typewriter standard of two hyphens (- -) or, with modern word processors, a character that is usually two or three hyphens in length (—). In either case, do not include spaces either before or after the dash. In modern word processing programs, two sequential hyphens are often interpreted as a dash with a single, longer character produced. CAUTION: Be careful not to overuse dashes, especially in formal writing where substitution for either commas or parentheses may be inappropriate, or where overuse could compromise the emphasis that dashes provide.

10.5.1 Indicating an Interruption or Break

This *dash* marks a sudden shift in thought from complimentary to threatening.

The professor said, "I'm sure you'll make a fine engineer—if you pass my class."

10.5.2 Emphasizing a Parenthetical Element

These *dashes* set off and emphasize a parenthetical comment.

The CEO said, "Invent a process—if you can—that will make this company grow."

10.5.3 Serving as an Introductory Connector

> This dash connects the introductory words, *fear and greed*, with the rest of the sentence that explains their context and meaning.

Fear and greed—that's what drives the stock market.

10.5.4 Used for Clarification

> For clarity, dashes are used here to set off a series of words in apposition with (i.e., placed beside to describe or explain) *steps*. Since the items in the series are themselves separated by commas, setting off the series with commas could be confusing.

Four steps—*developing, stopping, fixing,* and *washing*—were used in traditional, black and white, film processing.

10.6 Ellipsis

Use three spaced periods (each period separated by a single blank character space) when forming an ellipsis to indicate where information has been omitted in a quotation. If the omitted material comes at the end of a sentence, add a fourth period to the ellipsis as the terminal mark. Also, add a fourth period if the omitted material comes at the beginning of the sentence.

The observables of . . . electromagnetic waves are frequency, wavelength, and speed.

> An *ellipsis* indicates where the author has omitted part of the original material.

The professor noted, "X-rays and gamma rays are unique, and the process of their reflection is not well defined. . . ."

> The *ellipsis* at the end of the sentence has four dots. The first three dots show that some of the professor's words at the end of the quotation have been omitted. The fourth dot acts as a period at the end of the sentence.

10.7 **Exclamation** **Point** *(rarely, if ever,* *used in technical* *writing)*	Use an *exclamation point* when you want to provide unusually strong emphasis to a word (interjection), phrase, clause, or sentence. The exclamation point normally expresses surprise, conviction, disbelief, or strong emotion. By the way, be careful not to overuse exclamation points. Since they are reserved for *unusually* strong emphasis, their use must be unusual if they are to have the desired effect. In normal cases, use a comma instead of an exclamation point—but never add a comma or period after one.

10.7.1 Use with a Word (Interjection)

This **exclamation point** shows unusually strong emphasis for the interjection *wow*. In normal circumstances, a **comma** should be used instead.

Wow! I wish my car had that kind of acceleration.

Wow, I wish my car had that kind of acceleration.

Punctuation note: <u>Never</u> add a comma or period after an exclamation point.

Wow!, I wish my car had that kind of acceleration.
no comma

10.7.2 Use with a Sentence, Clause, or Phrase

An **exclamation point** is used to punctuate the end of an imperative sentence. Notice that **quotation marks** have also been used. Punctuation <u>other</u> than commas and periods may be used with exclamation points.

"The material is showing plastic deformation. Stop the test!"

Use hyphens to divide words at the end of a line
and to link compound words or word parts.

**10.8
Hyphen**

10.8.1 End of the Line

Words that must be split at the end of a line are
divided with a hyphen. This division always
occurs between syllables. Modern word processing software can automatically make these divisions, so hyphenation is no longer the major con-
cern it once was.

**Throughout the night and into the morning, we waited for the net-
work to degrade.**

> Here, *network* is divided
> between syllables by a
> hyphen.

10.8.2 Compound Words

Hyphens are often, but not always, used to
divide compound words acting together as a single unit. For example, generally, you would
hyphenate a compound adjective before a noun
but not after a noun. However, increasingly,
many compound adjectives are written as single
words, so you should consult a dictionary before
hyphenating if in doubt.

The *well-respected* chemist loved polymers.

> The compound adjective
> is hyphenated before the
> noun, but not after.

The chemist who loved polymers was *well respected*.

> Hyphens are <u>not used</u> when a compound
> adjective contains an adverb ending in
> *-ly*.

The highly respected chemist loved polymers.

10.8.3 Numbers and Fractions

Always use hyphens to spell out any number between twenty-one and ninety-nine no matter how and where it is used. When spelling out fractions, always use a hyphen between the two parts of a fraction, except when the fraction is being used as a noun.

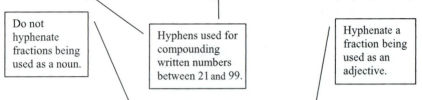

At the Twenty-One Club, he lost forty-five thousand dollars playing Blackjack.

Do not hyphenate fractions being used as a noun.

Hyphens used for compounding written numbers between 21 and 99.

Hyphenate a fraction being used as an adjective.

That represented one half of his net worth with only one-fifth liquidity remaining.

10.9
Parentheses

Use *parentheses* to enclose additional, parenthetical, and illustrative material within a sentence or in a separate sentence. Also, use parentheses to enclose figures or letters that are being used for enumeration.

10.9.1 To Enclose Material

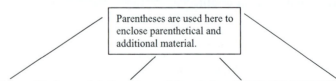

Parentheses are used here to enclose parenthetical and additional material.

High-level AM (amplitude modulation) is used for air band (109-136.975 MHz) voice communication.

10.9.2 To Enclose Enumeration Symbols

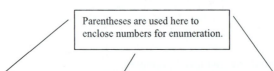

Parentheses are used here to enclose numbers for enumeration.

To log on, (1) enter your User ID, (2) enter your password, and (3) press <RETURN>.

Use a *period* to mark the end of a sentence where other marks of punctuation, like exclamation points or question marks, are not indicated. Also, use periods with most abbreviations and when creating ellipses to indicate omissions (see 10.6).

10.10 Period

10.10.1 End of a Sentence

Almost all declarative and imperative sentences end with a period. A period is also used to mark the end of an indirect question or a question presented purely as a courtesy in either a technical or business letter.

> A declarative sentence is terminated with a period.

Light can be seen as the periodic motion of both waves and particles.

> The imperative sentence is terminated with a period. A strong imperative sentence might use an exclamation point.

Read your physics assignment about wave and particle theory.

> An indirect question is punctuated with a period.

The boss wants to know if you plan on coming to work today.

> A question posed purely as a courtesy with no expectation of an answer, especially in business and technical letters, is terminated with a period.

Will you please contact me if you need more information.

10.10.2 Abbreviations

Most, but not all, abbreviations require periods. Do not use periods for abbreviations that stand for names of organizations or well-known agencies. Additionally, you may elect not to use periods for common abbreviations when the context of the writing clearly defines what the abbreviation represents.

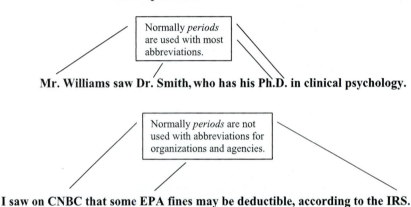

Normally *periods* are used with most abbreviations.

Mr. Williams saw Dr. Smith, who has his Ph.D. in clinical psychology.

Normally *periods* are not used with abbreviations for organizations and agencies.

I saw on CNBC that some EPA fines may be deductible, according to the IRS.

Anyone working in the computer business knows about UNIX and operating systems. Periods are not used for either UNIX or OS.

Bell Laboratory invented the UNIX OS as its Universal Interactive Executive.

Use a *question mark* after any direct question. A question mark is considered a terminal mark of punctuation, so never use a period or comma in a sentence after a question mark. Also, a question totally contained within parentheses or quotation marks has its question mark included inside the parentheses or quotation marks; otherwise, the question mark goes outside.

**10.11
Question Mark**

Will you give me the money, or do I have to work for it?

A *question mark* is always used with a direct question.

Here the *question mark* is placed inside the closing quotation mark of the directly quoted question. Notice that even though the overall sentence is declarative, no period follows the question mark.

He asked his boss, "Will you give me what I'm worth?"

Each element of this series of questions has been punctuated with a question mark. Each element is a direct question, and each question is emphasized by this punctuation.

Will you give me a raise? Put me in a large office? Allow me to travel first class?

**10.12
Quotation
Marks**

Use quotation marks to set off all direct quotations as well as minor titles such as book chapters, magazine articles, or specific television episodes. (Use italics or underlining for major titles such as books, magazines, motion pictures, or television series.) Also use quotation marks to set off words or symbols being used in a special way. With few exceptions, quotation marks enclose periods and commas, while other marks are only enclosed if they are a part of the quotation itself.

Quotation marks setting off a direct quotation of what the teacher told the class.

The teacher told the class, "Becquerel discovered radioactivity in 1896."

Single quotation marks enclose the inner quotation.

Jennifer said, "My teacher said, 'Becquerel discovered radioactivity in 1896.' "

Double quotation marks enclose the outer and inner quotation.

The minor title, here a chapter's title, is enclosed in quotation marks.

The major title, the book's title, is in *italics* (or underlined).

"On the idea of time in physics" is a chapter in Einstein's book *Relativity*.

The symbols for the code are set off with quotation marks. Notice the period, normally enclosed within quotation marks, is outside of the marks to prevent confusion over, or errors in, the code itself.

In HTML, the code for a lowercase *a* is "&#97".

A semicolon functions either to join or separate parts of a sentence. You can use a *semicolon* as a "soft" coordinating conjunction to join clauses, phrases, or other parts of a sentence <u>that carry equal grammatical rank</u>. In other words, you can use a semicolon to join either two main clauses, two nouns, or two verbs—but not a main clause and dependent clause, and not a noun and verb. A semicolon must also precede a conjunctive adverb that joins two main clauses. You should also use semicolons to separate items in a series when at least one of the items already includes a comma.

10.13
Semicolon

10.13.1 Joining Function

The semicolon acts as a coordinating conjunction and joins two closely related main clauses.

The compressor finally started; the condenser began radiating heat.

The semicolon precedes a conjunctive adverb, which, in turn, joins two main clauses.

The compressor started; <u>consequently</u>, the condenser began radiating heat.

10.13.2 Separating Items in a Series

In this example, two *semicolons* separate three items in a series. Because the first item itself contains two commas, *semicolons* (<u>not</u> *commas*) must be used to separate all three items in the series.

Once the compressor started, <u>the condenser, which had been cold, radiated heat</u>;
Item #1

<u>the expansion valve began operating</u>; <u>and the evaporator absorbed heat.</u>
Item #2   *Item #3*

10.13.3 Incorrect Usage of Semicolons

> Since none of these items in a series has an internal comma, the items themselves must be separated with *commas*, not *semicolons*.

The radio frequency amplifier ⁄ , plate coil ⁄ , and load capacitor were contained in a high-voltage cage.

**10.14
Slash**

For technical writing purposes, *slashes* are used formally in English for only one reason: to indicate the existence of options. That being said, slashes are also used informally for many other purposes, including an often-used, informal proofreader symbol for "deletion," as shown in 10.13.3.

> *True* and *false* are the two possible outputs for processor.

Given any input, the processor will provide a true/false response.

> *Active* and *retired* are the two options or categories of recipients for the report.

The report should be sent to active/retired members as appropriate.

11

Final Thoughts

This book begins with the observation that "grammar is nothing more than a collection of the rules that we use for assembling words so that, together, they make sense and convey meaning." Making sense and conveying meaning are what technical writing is all about. Even if you are an intelligent, creative engineer or scientist with great potential for contributing to society and your own career, if you cannot effectively communicate what you know to those who need to know it, then what you know will not count for much. Your knowledge and hard work will contribute neither to the betterment of society nor to the enhancement of your career—at least not to the extent it could have.

The bottom line: There is no downside to using correct grammar. Proper use of the English language is a win-win situation—for you and everyone else. Correct grammar in technical writing is essential for the accurate and effective communication of precise, scientific information—the very information from which we weave the fabric of our society's most basic functions. For that reason alone, the demand will only increase for engineers and scientists who are able to communicate effectively about technology, and so will the opportunities for their success and self-fulfillment.

This book does not claim to be a complete guide for all of the complexities and nuances of English grammar. However, this book is designed to provide the basics of English grammar in a way that is accessible, usable, and helpful for the engineers and scientists we all rely on.

12

Glossary of Grammar and Usage

The following glossary provides brief definitions, examples, and explanations of the most common terms and concepts associated with English grammar and usage. This alphabetically listed section is designed to help you find quick answers to your grammar questions, or failing that, to point you in the right direction. Cross-references are provided to other entries in this glossary (shown in *italics*), to specific chapters and chapter sections of this book (shown as chapter or section numbers), or to external sources. If you still cannot find what you are looking for, be sure to check the comprehensive index at the end of the book.

abbreviation
A shortened form of a word or phrase.

Cat 5 cable = Category 5 cable

2 million *pixels* = 2 million picture elements

HVAC = heating, ventilation, and air conditioning

RMS = root mean square, remote manipulator system, record management services

ATM = Adobe Type Manager, anti-tank missile, asynchronous transfer mode, automated teller machine
[Over a period of time many abbreviations, like these, may replace the original word or phase in daily use. Also, as exemplified by *RMS* and *ATM,* many meanings within the context of science and engineering are possible for a single abbreviation.]

absolute adjective
A modifier that turns the noun it modifies into a specific, nonmodifiable state. See *adjective.* See also 4.2.6.

The dead *scorpion* found in the waveguide was tossed into the garbage.
[The scorpion cannot be *partially dead* or *really dead.* Dead is absolute.]

absolute phrase
Also called a *nominative absolute* or *sentence modifier.* A parenthetical phrase consisting of a noun and a participle that acts as an adverb and modifies the rest of the sentence. Note: an *absolute phrase* is not grammatically related to the rest of the sentence by a connective (such as a

conjunction). See *parenthetical, adverb, adverb phrase, connective.* See also 6.1.2, 6.1.5.

The external power having failed, the system continued on backup batteries.
[The introductory absolute phrase includes the noun *power* and the participle *having failed.* It modifies the entire sentence but is not joined to the sentence with a connective.]

active voice
The voice of a transitive verb that takes a direct object and follows the pattern: subject *(doer)*—verb *(action)*—object *(receiver).* See *passive voice, verb.* See also 5.5.2.

The system administrator *learned* a hard lesson.
[A transitive sentence following the pattern: *subject—verb—direct object.* Here, *administrator* is the subject, *learned* is the verb, and *lesson* is the direct object.]

adjectival clause
Also called an *adjective clause.* A subordinate clause used as an adjective in a sentence. See *clause.* See also 4.4.

The scientist *who wrote this report* is brilliant.
[The clause *who wrote this report* acts as an adjective modifying *scientist.*]

adjectival noun/pronoun
A noun or pronoun that also functions as an adjective in a sentence. See 4.1.1.

Elizabeth loves *her* digital recorder.
[The pronoun *her* refers to the antecedent *Elizabeth* while also acting as an adjective modifying *recorder.*]

adjective

The part of speech that modifies, describes, or limits a noun or pronoun. See *participle, absolute adjective, predicate adjective.* See also Chapter 4.

The *powerful* transmitter was operating at 80 percent when hit by lightning.
[The adjective *powerful* modifies the noun *transmitter.*]

adverb

The part of speech that modifies, describes, or limits a verb, adjective, phrase, clause, or another adverb. See Chapter 6.

The computer booted *rapidly*.
[The adverb *rapidly* describes <u>how</u> the computer booted.]

adverb clause

See *adverbial clause.*

adverbial

Any word or group of words used as an adverb in a sentence. See *adverb, adverbial clause, adverbial conjunction, adverbial phrase.* See also 6.1.

adverbial clause

Also called an *adverb clause.* A subordinate clause used as an adverb in a sentence. See *adverb.* See also 6.1.6.

***As I tested the device*, I discovered its ergonomic deficiencies.**
[*As I tested the device* modifies the main clause that follows.]

adverbial conjunction

Also called a *conjunctive adverb*. A conjunction that functions as an adverb while joining main clauses. See *adverb, conjunction*. See also 8.4, Table 12.2.

The wing passed the vibration test; however, we still had our doubts.
[The conjunctive adverb *however* joins two main clauses.]

adverbial phrase

Any phrase used as an adverb in a sentence. See *absolute phrase, adverbial, phrase, infinitive phrase, prepositional phrase*. See also 6.1.2.

On June 21, 2004, Anchorage had 24 hours of visible light.
[A prepositional adverbial phrase describing *when.*]

To be precise, on June 21, 2004, Anchorage had 19 hours and 22 minutes of actual daylight and 24 hours of visible light.
[An infinitive adverbial phrase has been added to the front of the sentence.]

agreement

The conformity in form between one word and another word when both words are related in a sentence. See 3.1.2, 5.2, 5.3.

The *crewmembers* took *their* positions in the cockpit.
[The plural pronoun *their* conforms or agrees in person and number with its antecedent *crewmembers.*]

antecedent

Also called a *referent*. The word or group of words to which a pronoun refers. See *agreement*. See also 3.1.

The *xerographic plate* had *its* surface ionized by a high-voltage electrode.
[The pronoun *its* refers to the *xerography plate,* which is the antecedent.]

apostrophe

A mark of punctuation (') that indicates possession for a noun or substitutes for missing letters in a contraction. See 10.1.

She *can't* open the anechoic chamber.
[In the contraction *can't,* the apostrophe replaces two letters in *can̲n̲ot.*]

appositive

A noun or noun phrase placed beside another noun to describe, explain, or identify that noun. See 2.7.

Stephen, the *database administrator*, is always on call.
[The appositive *database administrator* describes and identifies *Stephen.*]

article

Also called a *determiner.* A special type of adjective that signals the presence of a noun that is just ahead in the sentence. The articles *a, an,* and *the* always precede a noun. See 4.3.

An unexpected cold air advection interacted with *a* pool of warm air causing *the* thunderstorms to intensify rapidly.
[The three articles—*a, an,* and *the*—are used in the sentence above.]

auxiliary verb

Also called a *helping verb*. A word paired with a main verb in a verb phrase to indicate tense, voice, mood, person, or number. See *participle, present participle, past participle.*

Once we *had fabricated* the chip successfully, the instructor was happy.
[The auxiliary verb *had* is paired with the past tense of *to fabricate.*]

brackets

Paired marks of punctuation ([ ]) used to set off an explanation or an editorial correction within quoted material. Can also be used to indicate a parenthetical level within a set of parentheses. See *parentheses, parenthetical.* See also 10.2.

The professor said, "If you move fast enough *[near the speed of light],* time dilation will occur."
[The bracketed material explains what *fast enough* means and is not part of the quotation.]

case

The category of a noun or pronoun that indicates its function in a sentence. Subjects are in the *subjective* or *nominative* case, objects are in the *objective case,* and possession is shown with the *possessive* case. See *subjective case, objective case, possessive case.* See also 2.4.

Her computer ran the simulation easily.
[The pronoun *her* is possessive, the noun *computer* is subjective, and the noun *simulation* is objective.]

clause
A group of related words in a sentence that contains a subject and a predicate. See *main clause, noun clause, subordinate clause.* See also 1.4.1.

The radar indicated a tornado vortex signature, and a warning was issued.
[Two main clauses are joined by the coordinating conjunction *and.*]

collective noun
A singular noun that often has a plural meaning and refers to a group or *collection* of persons, places, things, concepts, etc. See also 2.2.

This *team* of chip designers is one of the best I have seen.
[The collective noun *team* is singular.]

colon
A mark of punctuation (:) used to introduce and emphasize additional material. Also used in references and source documentation. See 10.3.

Xerography includes the following six steps: *charging, exposing, developing, transferring, fusing,* and *cleaning.*
[The colon is used here to introduce and emphasize the six steps that follow.]

comma
A mark of punctuation (,) that indicates brief pauses or inflection changes in a sentence. See 10.4.

I screamed, she pointed, but no one seemed to care.
[Commas separate three clauses in a series, indicating natural pause points.]

common noun
A noun that does <u>not</u> name a specific person, animal, place, thing, concept, or quality. See *proper nouns*. See also 2.3.

comparative levels
A conceptual framework for describing three comparative levels of either quality or amount (either negative or positive) for adverbs and most adjectives. The levels are described as *standard, comparative,* and *superlative.* See 4.5, 6.2.

Jim was *motivated*, Jane was *more motivated* than Jim, and Sam was the *most motivated* in the class.
[In this example, *motivated* is standard; *more motivated* is comparative of two people, <u>Jane and Jim</u>; and *most motivated* is the superlative of <u>Jim and everyone else</u> in the class.]

complement
One or more words that complete the sense of a subject, verb, or object. See 2.4.

The engineer knew *nothing about the device*.
[In this example, *nothing about the device* is the complement of the verb *knew*.]

complex sentence
A sentence that contains a main clause and a subordinate clause. See 1.4.3.

The computer was booted although the network was down.
[The subordinate clause *although the network was down* is dependent on the main clause *the computer was booted.*]

compound adjective

An adjective composed of two or more parts, each of which describes or limits a noun in some way. See *adjective*. See also 10.8.2.

He served an *elegant, formal* meal to his staff of design engineers.

[*Elegant* and *formal* both modify *meal*. *Elegant* says that the meal was pleasing to the eye, while *formal* says that it adhered to prescriptive and ceremonial form.]

compound adverb

A compound structure involving an adverb paired with, and modifying, another adverb. See *adverb*. See also 6.3.

The program was structured *very badly*.

[The adverb *very* modifies a second adverb *badly*.]

compound-complex sentence

A sentence that contains two main clauses and one or more subordinate clauses. See 1.4.4.

The server failed, and the network went down because no one paid attention.

[This sentence contains a main clause about the server failing, another main clause about the network going down, and a subordinate clause blaming inattention as the cause.]

compound sentence

A sentence that contains two main clauses. See 1.4.2.

The cylinder head cracked, and the engine stalled.

[The main clause about the cylinder head cracking is connected to another main clause about the engine stalling.]

compound subject
A subject consisting of two or more words. See *subject*.

> **Bill and Mary are the system administrators.**
> [The compound subject *Bill and Mary* is plural and takes the plural verb *are*.]

conjugation
A listing of the inflected forms of a verb to show tense, person, and number (Table 12.1). Conjugations can also include mood and voice. See *tense, person, number*. See also 5.2, 5.3.

Table 12.1 Conjugation of the present tense of the verb *to be*

Person	Singular	Plural
First	I am	we are
Second	you are	you are
Third	he/she/it is	they are

conjunction
The part of speech used to connect words, phrases, clauses, and sentences. See *clause, phrase, sentence, coordinating conjunction, correlative conjunction, subordinate conjunction, adverbial conjunction* (Table 12.2). See also Chapter 8.

Table 12.2 Types of conjunctions

Example	Type
Bill *and* Mary work together.	**coordinating**
Both Bill *and* Mary have children.	**correlative**
Because Bill and Mary work together, they are friends.	**subordinating**
Bill and Mary work together; *consequently,* they are friends.	**adverbial**

conjunctive adverb
See *adverbial conjunction.*

connective
A word or phrase that joins, links, and relates other words, phrases, clauses, or sentences. Common connectives include conjunctions and conjunctive adverbs. See *adverbial conjunction, conjunction.* See also Chapters 6 and 8.

The engine was expensive; *however,* its performance was excellent.
[In this sentence, *however* is the connective that joins two main clauses.]

coordinating conjunction
A conjunction used to connect words, phrases, clauses, and sentences of equal rank or importance. See Table 12.2. See also 8.1.

The intake valve closed, *and* the piston compressed the mixture.
[The coordinating conjunction *and* connects two main clauses. Coordinating conjunctions that link main clauses are always preceded by a comma.]

correlative conjunction
A pair of conjunctions used to connect parallel structures. See Table 12.2. See also 8.2.

Either a scientist *or* an engineer should fill this position.
[*Either/or* connects *scientist/engineer* in parallel fashion.]

dangling modifier
Sometimes called a *dangling participle* or *misplaced modifier.* A modifier, usually a participle phrase or adverb clause, that does not clearly refer to another word or phrase in a sentence because of its placement in the sentence. This phenomenon is called a *syntactical ambiguity* because the ambiguous meaning relates not to the words being used, but rather to their physical misplacement in a sentence. See 4.1.3.

> **Corrupted by a power spike, the operator shut down the process.**
> [This sentence seems to say that the operator was corrupted by the power spike. Because of its placement, the introductory participial phrase *corrupted by a power spike* modifies *operator,* when it should modify *process.* The process was corrupted. The operator may have been corrupted by money, lust, or greed, but not a power spike.]

dash
A mark of punctuation (—) that produces a sharp break in a sentence to indicate an interruption or change in thought. It is also used to emphasize parenthetical elements. See 10.5.

> **He showed up for work half asleep—the night shift had called him at home at 2 a.m.—yet he quickly dealt with the emergency.**
> [The night shift's calling him at home represents a complete change in thought and is set off with dashes.]

declarative mood
See *mood*.

demonstrative pronoun
A pronoun that points to a noun and clarifies which group, person, place, thing, or idea is under discussion. The demonstrative pronouns are *this, that, these,* and *those.* See *pronoun.* See also 3.2.7.

Those are the concepts of quantum physics that I do not understand.
[*Those* points to specific ideas that should have been mentioned prior to this sentence.]

dependent clause
See *subordinate clause*.

descriptive adjective
An adjective that describes the attributes of the noun it modifies. See *adjective.* See also 4.2.1.

The *fast* processor compensated for the *inefficient* program.
[The adjective *fast* describes the speed of the processor, while *inefficient* describes the poor design of the program.]

determiner
See *article*.

diagram
A graphical description of a sentence that shows the interrelationship of its parts (Figure 12.1). See 1.4.

ESL

An acronym for English as a Second Language.

exclamation point

A punctuation mark (**!**) that provides strong emphasis. See *interjection*. See also 10.7.

expletive

The words *there* or *it* when used to lead off a sentence without adding any additional meaning to the sentence.

There was much interest in the new fabrication process.

[*There* adds no additional meaning. The sentence could be rewritten: *Much interest existed in the new fabrication process.*]

form

The structure of a verb that determines whether it requires an object to receive its action. *Transitive* forms require an object, while *intransitive* forms do not. See *transitive, intransitive*. See also 5.5.

The test device destroyed the sample.

[The <u>transitive</u> verb *destroyed* takes the direct object *sample.*]

The test device malfunctioned.

[The <u>intransitive</u> verb *malfunctioned* requires no object.]

gerund

A verb used as a noun. See *verbal*. See also 2.4.1.

The *heating* of a moist atmosphere can destabilize it.

[The gerund *heating* is the subject of this sentence.]

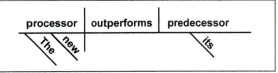

Figure 12.1
Traditional sentence diagramming.

[In traditional diagramming (above), the subject is to the left of the full vertical line, the predicate is to the right of this line, and the verb is separated from the rest of the predicate by the half vertical line. The diagonal descending lines identify modifiers.]

direct object

A noun or noun substitute that receives the action of the verb and is directly affected by that action. See *object*. See also 2.4.

The fuel cell quickly energized the *circuit*.

[The direct object *circuit* receives the verb's action.]

direct quotation

A representation of the exact words that were written or spoken by someone. Direct quotations are usually enclosed in quotation marks. See *indirect quotation*. See also 10.12.

The chief scientist said, *"Get out of my office."*

[The chief scientist said exactly the words within the quotation marks.]

draft
Any written material that is not complete and in final form.

editing
The iterative process designed to refine, correct, redirect, or enhance written material.

ellipsis
A sequence of three dots (. . .) indicating an omission in a quotation. See 10.6.

The technician said, "The spectrometer is broken and not working."

The technician said, "The spectrometer is . . . not working."
[An ellipsis is used to indicate the omission of words.]

The technician said, "The spectrometer is broken. . . ."
[An ellipsis is used with a period at the end of a sentence resulting in 4 dots—3 for the ellipsis and one for the period.]

elliptical construction
A construction in which words are omitted but are clearly understood to be present.

My computer is smarter than I [am smart].
[Because *am smart* is understood, it is omitted after *I*.]

*English: British vs. American**
The differences between American English and British English are relatively minor and usually

*For more detailed information, see "Proper Treatment: British vs. American." Internet: http://www.digitas .harvard.edu/cgi-bin/wiki/ken/BritishvsAmerican.

---

cause few serious problems. The differences involve spelling (Table

Table 12.3 Examples of regular spelling d

American	British
-or (color)	-our (colour)
-er (center)	-re (centre)
-g (analog)	-gue (analogue)
-ize (realize)	-ise (realise)
-ction (deflection)	-xion (deflexion)
-e (medieval)	-ae (mediaeval)
-e (fetus)	-oe (foetus)
-l (modeling)	-ll (modelling)
-g (aging)	-ge (ageing)
-se (license)	-ce (licence)
-m (program)	-mme (programme

Other differences exist in basic gr example, some collective, singular n ican English, such as *team,* are oft plural nouns in British English. ences also exist in the meaning ar certain words, as demonstrated in that follow.

Americans *watch* their step w into an *elevator*, while the B their step when getting into *a*
[As used here in British English, "watch" and *lift* means "elevator."

The *Mississippi River* is in Am the *River Thames* is in the Gr
[The British usually put the wor not last.]

*For more detailed information, see the *Engli Dictionary*. Internet: http://english2american.

grammar
The collection of rules used for assembling words so they make sense and convey meaning. See 1.1.

helping verb
See *auxiliary verb*.

hyphen
A mark of punctuation used to divide words at the end of a sentence and to connect compound words acting together as a single unit. See 10.8.

Magnetic levitation (MAGLEV) trains often rely on *low-temperature super-conductors* for propulsion.
[The compound expression *low-temperature* is hyphenated, as is the line break in the word *superconductors*.]

idiom
An expression that means something other than its literal meaning. See 7.2.

The technician said, "Please *give me a hand*."
[In this example, *give me a hand* is idiomatic. The technician is asking for assistance, not a body part.]

imperative mood
See *mood*.

indefinite pronouns
A pronoun that does not require an antecedent. See *pronoun*. See also 3.2.4.

If we don't upgrade our software, *nothing* will get done.
[The pronoun *nothing* does not require an antecedent.]

independent clause

A synonym for *main clause*. See *main clause*.

indirect object

A noun or noun substitute that shows to whom or what the verb's action flows. See *object*. See also 2.4.

He gave the faulty *engine* a complete overhaul.
[Here, *engine* is the indirect object. The engine <u>receives</u> the overhaul. *Overhaul* is the direct object.]

indirect quotation

A reworded direct quotation that accurately reports the sense and order of the original written or spoken words. See *quotation*. See also 10.12.

The chief scientist said *to get out of her office*.
[An indirect quote of the chief scientist who actually said, "Get out of my office."]

infinitive

A verb used as a noun, adjective, or adverb. An infinitive is composed of *to* + the base form of the verb called the *verb stem*. See *infinitive phrase, verb, verbal, verb stem*.

He loved *to make* money.
[In this example, *to make* is an infinitive.]

infinitive phrase

A phrase containing an infinitive that is being used as a noun, adjective, or adverb. See *infinitive*. See also 4.1.2, 6.1.2, 7.1.

He loved *to make money*.
[The infinitive phrase *to make money* is a noun acting as the direct object.]

inflection
In written English, a change in the form of a noun, pronoun, verb, or modifier to alter its meaning or its grammatical relationship with something else in the sentence. Inflections are normally used to indicate tense, person, number, or mood. See *tense, person, number, mood, conjugation.* See also Chapter 5.

intensive pronoun
A pronoun used to emphasize another word in a sentence. See *pronoun.* See also 3.2.3.

He *himself* did all the work.
[In this sentence, *himself* intensifies the fact that he and only he did the work.]

interjection
The part of speech used either as an exclamation or to show surprise, emotion, or impact. See Chapter 9.

***Oh no!* I'm getting older and my hair is falling out.**
[The interjection may be one or more words used with or without an exclamation point.]

interrogative mood
See *mood.*

interrogative pronoun
A pronoun normally placed at the beginning of a sentence to initiate a question. See *pronoun.* See also 3.2.6.

***Who* will replace the equipment that was damaged?**
[*Who* at the start of this sentence signals a question.]

intransitive verb
A verb that does not take an object. See *transitive verb*. *See* also 5.5.1.

The milling machine *functioned* perfectly.
[The verb *functioned* does not take an object and is intransitive.]

irregular noun
A noun that does not follow standard rules for pluralizing. See *regular noun*. See also 2.2.1.

The plural of *mouse* is *mice*.
[More than one mouse is *mice,* not *mouses.*]

irregular verb
A verb that does not form its past and past participle by adding *-d* or *-ed* to the verb stem. See *tense, regular verb*. See also 5.4 and Table 5.3.

The engineers *swim* today while the scientists *swam* yesterday.
[*Swim* is the present form, while *swam* is the past of this irregular verb.]

limiting adjective
An adjective that in some way limits the noun it modifies. See *adjective*. See also 4.2.2.

Submarines communicate at depth using *extremely low frequency* (ELF) radio waves.
[By convention, the adjective *extremely low frequency* limits *radio waves* to a standard frequency range of 3 Hz–30 Hz.]

linking verb
A form of the verb *to be* (is, are, was, were) or a sense verb (seem, look, appear, smell) that directly relates the subject to its complement. See 5. 5.1.1.

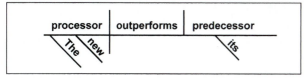

Figure 12.1
Traditional sentence diagramming.

[In traditional diagramming (above), the subject is to the left of the full vertical line, the predicate is to the right of this line, and the verb is separated from the rest of the predicate by the half vertical line. The diagonal descending lines identify modifiers.]

direct object
A noun or noun substitute that receives the action of the verb and is directly affected by that action. See *object*. See also 2.4.

The fuel cell quickly energized the *circuit*.
[The direct object *circuit* receives the verb's action.]

direct quotation
A representation of the exact words that were written or spoken by someone. Direct quotations are usually enclosed in quotation marks. See *indirect quotation*. See also 10.12.

The chief scientist said, *"Get out of my office."*
[The chief scientist said exactly the words within the quotation marks.]

draft
Any written material that is not complete and in final form.

editing
The iterative process designed to refine, correct, redirect, or enhance written material.

ellipsis
A sequence of three dots (. . .) indicating an omission in a quotation. See 10.6.

The technician said, "The spectrometer is broken and not working."

The technician said, "The spectrometer is . . . not working."
[An ellipsis is used to indicate the omission of words.]

The technician said, "The spectrometer is broken. . . ."
[An ellipsis is used with a period at the end of a sentence resulting in 4 dots—3 for the ellipsis and one for the period.]

elliptical construction
A construction in which words are omitted but are clearly understood to be present.

My computer is smarter than I [am smart].
[Because *am smart* is understood, it is omitted after *I*.]

English: British vs. American*
The differences between American English and British English are relatively minor and usually

---

*For more detailed information, see "Proper Treatment: British vs. American." Internet: http://www.digitas .harvard.edu/cgi-bin/wiki/ken/BritishvsAmerican.

cause few serious problems. The most obvious differences involve spelling (Table 12.3).

Table 12.3 Examples of regular spelling differences

American	British
-or (color)	-our (colour)
-er (center)	-re (centre)
-g (analog)	-gue (analogue)
-ize (realize)	-ise (realise)
-ction (deflection)	-xion (deflexion)
-e (medieval)	-ae (mediaeval)
-e (fetus)	-oe (foetus)
-l (modeling)	-ll (modelling)
-g (aging)	-ge (ageing)
-se (license)	-ce (licence)
-m (program)	-mme (programme)

Other differences exist in basic grammar.* For example, some collective, singular nouns in American English, such as *team,* are often treated as plural nouns in British English. Other differences also exist in the meaning and/or order of certain words, as demonstrated in the examples that follow.

Americans *watch* their step when getting into an *elevator*, while the British *mind* their step when getting into *a lift*.
[As used here in British English, *mind* means "watch" and *lift* means "elevator."]

The *Mississippi River* is in America, while the *River Thames* is in the Great Britain.
[The British usually put the word *river* first, not last.]

---

*For more detailed information, see the *English-to-American Dictionary*. Internet: http://english2american.com.

ESL
An acronym for English as a Second Language.

exclamation point
A punctuation mark (**!**) that provides strong emphasis. See *interjection*. See also 10.7.

expletive
The words *there* or *it* when used to lead off a sentence without adding any additional meaning to the sentence.

> **There was much interest in the new fabrication process.**
> [*There* adds no additional meaning. The sentence could be rewritten: *Much interest existed in the new fabrication process.*]

form
The structure of a verb that determines whether it requires an object to receive its action. *Transitive* forms require an object, while *intransitive* forms do not. See *transitive, intransitive*. See also 5.5.

> **The test device destroyed the sample.**
> [The <u>transitive</u> verb *destroyed* takes the direct object *sample*.]

> **The test device malfunctioned.**
> [The <u>intransitive</u> verb *malfunctioned* requires no object.]

gerund
A verb used as a noun. See *verbal*. See also 2.4.1.

> **The *heating* of a moist atmosphere can destabilize it.**
> [The gerund *heating* is the subject of this sentence.]

grammar
The collection of rules used for assembling words so they make sense and convey meaning. See 1.1.

helping verb
See *auxiliary verb*.

hyphen
A mark of punctuation used to divide words at the end of a sentence and to connect compound words acting together as a single unit. See 10.8.

Magnetic levitation (MAGLEV) trains often rely on *low-temperature super-conductors* for propulsion.
[The compound expression *low-temperature* is hyphenated, as is the line break in the word *superconductors*.]

idiom
An expression that means something other than its literal meaning. See 7.2.

The technician said, "Please *give me a hand*."
[In this example, *give me a hand* is idiomatic. The technician is asking for assistance, not a body part.]

imperative mood
See *mood*.

indefinite pronouns
A pronoun that does not require an antecedent. See *pronoun*. See also 3.2.4.

If we don't upgrade our software, *nothing* will get done.
[The pronoun *nothing* does not require an antecedent.]

independent clause
A synonym for *main clause*. See *main clause*.

indirect object
A noun or noun substitute that shows to whom or what the verb's action flows. See *object*. See also 2.4.

He gave the faulty *engine* a complete overhaul.
[Here, *engine* is the indirect object. The engine <u>receives</u> the overhaul. *Overhaul* is the direct object.]

indirect quotation
A reworded direct quotation that accurately reports the sense and order of the original written or spoken words. See *quotation*. See also 10.12.

The chief scientist said *to get out of her office*.
[An indirect quote of the chief scientist who actually said, "Get out of my office."]

infinitive
A verb used as a noun, adjective, or adverb. An infinitive is composed of *to* + the base form of the verb called the *verb stem*. See *infinitive phrase, verb, verbal, verb stem*.

He loved *to make* money.
[In this example, *to make* is an infinitive.]

infinitive phrase
A phrase containing an infinitive that is being used as a noun, adjective, or adverb. See *infinitive*. See also 4.1.2, 6.1.2, 7.1.

He loved *to make money*.
[The infinitive phrase *to make money* is a noun acting as the direct object.]

inflection
In written English, a change in the form of a noun, pronoun, verb, or modifier to alter its meaning or its grammatical relationship with something else in the sentence. Inflections are normally used to indicate tense, person, number, or mood. See *tense, person, number, mood, conjugation.* See also Chapter 5.

intensive pronoun
A pronoun used to emphasize another word in a sentence. See *pronoun.* See also 3.2.3.

> He *himself* did all the work.
> [In this sentence, *himself* intensifies the fact that he and only he did the work.]

interjection
The part of speech used either as an exclamation or to show surprise, emotion, or impact. See Chapter 9.

> *Oh no!* I'm getting older and my hair is falling out.
> [The interjection may be one or more words used with or without an exclamation point.]

interrogative mood
See *mood.*

interrogative pronoun
A pronoun normally placed at the beginning of a sentence to initiate a question. See *pronoun.* See also 3.2.6.

> *Who* will replace the equipment that was damaged?
> [*Who* at the start of this sentence signals a question.]

intransitive verb

A verb that does not take an object. See *transitive verb*. *See* also 5.5.1.

The milling machine *functioned* perfectly.
[The verb *functioned* does not take an object and is intransitive.]

irregular noun

A noun that does not follow standard rules for pluralizing. See *regular noun*. See also 2.2.1.

The plural of *mouse* is *mice*.
[More than one mouse is *mice,* not *mouses.*]

irregular verb

A verb that does not form its past and past participle by adding *-d* or *-ed* to the verb stem. See *tense, regular verb*. See also 5.4 and Table 5.3.

The engineers *swim* today while the scientists *swam* yesterday.
[*Swim* is the present form, while *swam* is the past of this irregular verb.]

limiting adjective

An adjective that in some way limits the noun it modifies. See *adjective*. See also 4.2.2.

Submarines communicate at depth using *extremely low frequency* (ELF) radio waves.
[By convention, the adjective *extremely low frequency* limits *radio waves* to a standard frequency range of 3 Hz–30 Hz.]

linking verb

A form of the verb *to be* (is, are, was, were) or a sense verb (seem, look, appear, smell) that directly relates the subject to its complement. See 5. 5.1.1.

He *is* dirty, and his clothes *smell* bad.
[The verb *is* relates the predicate adjective *dirty* to the pronoun *he,* and the verb *smell* relates the predicate adjective *bad* to the noun *clothes.*]

main clause
A clause that is not dependent on any other part of a sentence. See *clause.* See also 1.4.

The crystal vibrated when excitation occurred.
[*The crystal vibrated* expresses a totally independent thought.]

mass noun
Also called an *uncountable noun.* A noun that is both singular and plural and often refers to "how much" rather than "how many" for practical reasons. See 2.2.

How much *insight* did the analyst have?
[The noun *insight* is not something that can be readily counted. Also note that *insight* is both singular and plural.]

misplaced modifier
See *dangling modifier.*

mode
See *mood.*

modifier
Word or words acting as an adjective or adverb to limit, qualify, or describe another word or group of words.

This *critical* task was assigned to our *top-rated* office.
[In this sentence *critical* modifies the noun *task* and *top-rated* modifies the noun *office.*]

mood

Also called *mode*. The characteristic of a verb that indicates the writer's intent to make a statement (declarative), ask a question (interrogative), give a command (imperative), or create a conditional statement or one contrary to fact (subjunctive). (Table 12.4) See 5.6.

Table 12.4 Moods of English

Example	Mood
I *want* you to give me a better video display.	Declarative makes a statement.
Will you give me a better video display?	Interrogative* asks a question.
Give me a better video display.	Imperative gives a command.
If I were more important, you would have given me a better video display.	Subjunctive provides a statement that is conditional and hypothetical.

*Many scholars no longer consider the interrogative mood a part of Standard English grammar and now include it under the declarative mood. However, for consistency with other languages (and because the author had to learn it when he was young), the interrogative mood lives on in this book.

nominative case

A synonym for the *subjective case*. See *case*. See 2.4.

nonrestrictive

A word, phrase, or clause that is not essential to the meaning of a sentence. See *restrictive*. See also 10.4.4.

The transmission electron microscope, *I might mention,* is still not working.

[The clause *I might mention* is nonrestrictive—it is not essential to the meaning of the sentence.]

noun

Words that name persons, animals, places, things, concepts, qualities, or actions. See *proper noun, common noun*. See also Chapter 2.

noun clause
A clause that functions as a noun in a sentence. See *clause, noun phrase*. See also 2.8.

> **Whatever we fabricate has to be better than the current device.**
> [*Whatever we fabricate* is a noun clause acting as the subject in this sentence.]

noun phrase
A group of words without either a subject or a verb that functions as a noun in a sentence.

> **Programming a computer can be challenging.**
> [The noun phrase *programming a computer* acts as the subject of the sentence.]

number
The form of a verb that denotes whether the subject includes one (singular)—or more than one (plural)—person, place, thing, concept, quality or action. See 5.3.

> **One workstation *is* good, but several workstations *are* better.**
> [Here, one *workstation* takes the singular form *is,* while several *workstations* take the plural form *are.*]

object
A noun or noun substitute that receives the action of a transitive verb. See *direct object, indirect object*. See also 5.5.

> **He calibrated the *monitor*.**
> [The direct object *monitor* receives the action of the verb *calibrated.*]

objective case
The case that includes all objects. See *subjective case, possessive case*. See also 2.4.

order
The arrangement of words in a sentence. See *dangling modifier.* See also 1.4, 4.1.3, 4.2.4, 6.4.

parenthesis
A mark of punctuation (( )) used to enclose parenthetical material whether or not that material forms a sentence. See *parenthetical.* See also 10.9.

parenthetical
A qualifying, amplifying, or explanatory word, phrase, or clause set off from the rest of the sentence by parentheses—or by commas as well as dashes—as if the material were set off by parentheses. Parenthetical elements are always nonrestrictive. See *nonrestrictive, absolute phrase, parenthesis, comma, dash.* See also 10.4.4, 10.9.

The system was broken again (*it never did work well*).
[The parenthetical material is nonrestrictive and is set off with parentheses.]

The system, *which no one really liked,* once again failed at a critical time.
[The parenthetical material is nonrestrictive and is set off with commas.]

participle
A verb used as an adjective. The present participle is normally the *-ing* form of the verb, while the past participle is normally the *-ed* form. See 4.1.2. See also *adjective, verbal.*

The *freezing* rain caused significant accumulations of ice.
[The present participle of the verb *to freeze* is an adjective modifying *rain.*]

parts of speech
The eight classes of words that indicate different functions in a sentence. See *noun, pronoun, adjective, verb, adverb, preposition, conjunction, interjection.* See also 1.3.

passive voice
The voice of a transitive verb in which the receiver of the verb's action precedes the verb in a sentence. A passive verb contains a form of the verb *to be* and the past participle of the verb. When passive voice is used, the receiver of the action becomes the subject, and the doer of the action follows the verb—although the doer need not be included. The structure of a passive sentence is *receiver-action-doer* or *receiver-action.* Passive voice is the opposite of active voice and effectively swaps the traditional functions of the subject and object. See *past participle, verb, active voice.* See also 5.5.2.

The operator partitioned the disk.
[An active sentence with the subject *operator* preceding the verb *partitioned,* and the direct object *disk* following and receiving the verb's action.]

The disk *was partitioned* by the operator.
[Passive sentence with the doer following the verb. Notice that the subject is acted upon rather than being the actor.]

The disk *was partitioned.*
[Passive sentence with the doer omitted. In a passive sentence like this, the receiver of the verb's action effectively serves as the subject.]
Note: Passive voice is often used in scientific and technical writing when the doer of the action is not important and the receiver of the action is the main focus.

past participle

The past tense of a verb used as an adjective. Past participles are normally formed using the *-ed* form of regular verbs, or the *-en, -n, -t* forms of irregular verbs. See *participle, passive voice.* See also 4.1.2.

Exhausted from studying, Joe tripped on a step and later found that his wrist was broken.
[These past participles were formed from *to exhaust,* a regular verb, and *to break,* an irregular verb.]

period

A mark of punctuation (.) that usually indicates the end of a declarative, imperative, or subjunctive sentence. Also used with most abbreviations and ellipses. See *mood, abbreviation, ellipses.* See also 10.10.

The new lab chief had his Ph.D. in Chemistry.
[Periods are used here with the abbreviation and at the end of the sentence.]

person

The form of a verb or noun that indicates whether the subject is speaking, being spoken to, or being spoken about. (Table 12.5). See also 5.3.

Table 12.5 Person in English (singular)

First person	I am in charge.	(The person speaking)
Second person	You are in charge.	(The person spoken to)
Third person	William is in charge.	(The person spoken about)

personal pronouns

Pronouns that substitute for people or things. See *pronoun*. See also 3.2.1.

They liked *his* **plasma display and** *her* **furniture, while** *I* **just liked** *her*.
[Here, *they, his, her,* and *I* are examples of personal pronoun use.]

phrase

A group of words that does not contain either a subject or a complete verb. See *absolute phrase, adverbial phrase, infinitive phrase, noun phrase, prepositional phrase*. See also 4.1.2, 6.1.2, 7.1.

Within the radiation acquired dose (RAD)
[This is an example of a prepositional phrase. It does not contain a verb.]

possessive case

The case that shows possession, ownership, control, or custody. See *subjective case, objective case*. See also 2.4.

The *company's* **laboratory is** *his* **area of responsibility.**
[In this example, *company's* shows ownership of the laboratory, and *his* shows possession of the area of responsibility.]

predicate

The part of a sentence that asks or asserts something about the subject in a sentence. See *predicate adjective, predicate noun*. See also 1.4, 5.5.1.1.

The phase shift *occurred without warning*.
[The predicate *occurred without warning* asserts something about *the phase shift*.]

predicate adjective

An adjective that follows a linking verb and modifies the subject. See *predicate noun, linking verb*. See also 4.2.5, 5.5.1.1.

Emily is *happy*, while Joe feels *sad*.

[The predicate adjectives *happy* and *sad* describe their respective subjects *Emily* and *Joe*.]

predicate noun/pronoun

Also called a *predicate nominative*. A noun or pronoun that follows a linking verb and modifies the subject. A predicate nominative is in the same case as the subject. Predicate nouns are in the *subjective* or *nominative* case, never the *possessive* or *objective*. See *predicate adjective, linking verb*. See also 5.5.1.1.

The new technician is *James*.

[The predicate nominative *James* is linked to, and equates with, *technician*.]

preposition

The part of speech used to show the relationship of a noun or noun substitute with another word in the sentence. See Chapter 7.

The old engineer is *in* the mood *for* love.

[Here, the preposition *in* relates the noun *mood* to the noun *engineer,* and the preposition *for* relates the noun *love* to the noun *mood.*]

prepositional phrase

A phrase that results when a preposition is combined with a noun or pronoun. See *preposition*. See also Chapter 7.

present participle
The form of a verb that ends in *-ing*. The present participle is typically used as a gerund, as part of a verb phrase, or as an adjective. See *participle*. See also 4.1.2.

> **Calibrating the sensor pack required many hours.**
> [*Calibrating* is a gerund acting as the subject of the noun phrase, *calibrating the sensor pack*. The noun phrase, in turn, acts as the subject of the sentence.]

> **The technicican *calibrating the sensor pack* has been well trained.**
> [*Calibrating the sensor pack* forms a verb phrase that modifies *technician*.]

> **We completed our calibrating procedures before the deadline.**
> [Here, *calibrating* acts as an adjective modifying *procedures*.]

principal clause
A synonym for *main clause* or *independent clause*. See *main clause*.

progressive
A verb tense form showing that the action occurs progressively. See 5.2, Table 5.1.

> **The computer *is* successfully *running* the program.**
> [The action of program execution is progressing successfully.]

pronoun
A generic word that takes the place of a noun in a sentence and functions in the sentence exactly as the noun it replaces. See *personal, reflexive, intensive, indefinite, interrogative, demonstrative, reciprocal,* and *relative pronouns*. See also 3.2.

proper adjective

An adjective derived from a proper noun. Note: proper adjectives are usually capitalized in a sentence, but in the same way the proper nouns from which they are derived are capitalized. For example, some proper nouns, especially trademarks, may not capitalize the first letter while capitalizing other letters. In this case, the proper adjective's capitalization must match precisely that of the proper noun. See *proper noun*. See also 4.2.3.

The *Mac OS/X* operating system is produced by Apple Computer, Inc.
[*Mac OS/X* is a proper adjective modifying *operating system.*]

He used the *eTrade Financial*® online investment resources.
[The proper adjective precisely matches the proper noun's irregular capitalization.]

proper noun

A noun that names a *specific* person, animal, place, thing, concept, or quality. See *common noun*. See also 2.3.

Windows XP is an operating system produced by Microsoft Corporation.
[*Windows XP* names a specific operating system.]

punctuation

The use of symbols, marks, and signs to separate or connect certain parts of a sentence in order to clarify meaning. See Chapter 10.

question mark

The mark of punctuation (**?**) used after any direct question and as the terminal mark of punctuation for an interrogative sentence. See *mood.* See also 10.11.

> **Will you recalibrate the motion sensors in the test vehicle?**
> [Here a question mark is used to terminate an interrogative sentence.]

quotation
See *direct quotation, indirect quotation, quotation marks.*

quotation marks
Paired marks of punctuation (" ") used to set off or enclose all direct quotations, minor titles, book chapters, magazine articles, or television episodes. Quotation marks are also used to set off words or symbols being treated in a special way. If a quotation occurs *within* a quotation, then double marks are used outside of single marks. See 10.12.

> **The applicant said: *"The boss said, 'Hire me and you won't regret it.' He told me to tell you that."***
> [Single and double quotation marks are used here to set off a direct quotation within a direct quotation.]

reciprocal pronoun
A pronoun that occurs in pairs to reinforce the relationship between two separate antecedents. See *pronoun.* See also 3.2.8.

> **Mary and Cindy help *each other* all the time.**
> [The pronouns *each* and *other* refer to Mary and Cindy reciprocally. Each helps the other.]

referent
See *antecedent.*

reflexive pronoun

A personal pronoun used to reflect the action from the verb back to the subject. See *pronoun*. See also 3.2.2.

They saved *themselves* with quick thinking.
[The pronoun *themselves* reflects the verb's action back to the subject *they*.]

regular noun

A noun that follows standard rules for pluralizing. See *irregular noun*. See also 2.2.

The plural for *computer* is *computers* and for *dish* is *dishes*.
[Standard rules pluralize nouns by adding *-s* or *-es*.]

regular verb

A verb that forms its past and past participle by adding *-d* or *-ed* to the verb stem. See *irregular verb*. See also 5.3.

The engineers *believed* their own research.
[The regular verb *to believe* forms its past participle by adding *-d*.]

relative pronoun

Also called a *subordinating pronoun* or a *subordinate-clause marker*. A pronoun used as a subordinator to introduce a subordinate or dependent clause. See *pronoun*. See also 3.2.5.

Please determine the graphics processor *that* we need.
[The relative pronoun *that* subordinates the clause *we need*.]

restrictive
A word, phrase, or clause that is essential to the meaning of a sentence. See 4.4, 10.4.4.

The fuel *purchased this morning* might be contaminated.
[The phrase is essential to the meaning of the sentence because only the fuel that was *purchased this morning* might be contaminated.]

semicolon
A mark of punctuation (;) used to join clauses or parts of a sentence of equal rank. Also used to separate items of a series when any one item has a comma included as an internal mark of punctuation, and to punctuate adverbial conjunctions. See *adverbial conjunction, comma.* See also 8.4, 10.13.

Here is what occurred: *the volcano erupted, as it had before; lava flowed down the western slope; and the town was evacuated.*
[Semicolons are used to separate items in a series because the first item has an internal comma.]

sense verb
A linking verb that involves sensory stimuli (feel, taste, look, sound, and smell). See *linking verb, predicate adjective.* See also 4.2.5, 5.5.1.1.

After being struck by lightning, the transformer *smelled* bad.
[The sense verb *to smell* links the *transformer* in an olfactory way to the predicate adjective *bad.*]

sentence
The basic unit of expression, containing a subject and predicate, that can stand alone as an independent thought. See 1.4.

sexist usage
The use of words inherently demeaning of men or women. Often involves the generic use of male pronouns in the third person singular to refer to male and female antecedents when such specificity is dysfunctional or inappropriate. See 3.2.1.1.

Everyone in the lab must do *his* part to help *mankind*.
[Both *his* and *mankind* refer to groups that contain women as well as men. To avoid gender specificity, this sentence might be rewritten to read, "All lab personnel must do their part to help humanity."]

simple sentence
A sentence that contains a single main clause. See *sentence, compound sentence, complex sentence,* and *compound-complex sentence.* See also 1.4.1.

The data became corrupted in transmission.
[This simple sentence contains only one subject and one predicate.]

slash
The mark of punctuation (/) used to indicate the existence of options. See 10.14.

The code will produce a *true/false* result.
[The slash indicates an option for the result.]

split infinitive
In an infinitive, the result of putting a word or words (usually an adverb) between *to* and the verb stem. Once frowned upon, *split infinitives* are now acceptable if splitting them makes the sentence read better or clarifies meaning. See 6.1.4.

> To *actually* make money is the goal of many.
> [The adverb *actually* splits the infinitive *to make*. In this case, the split infinitive is acceptable because not doing so would have resulted in *actually to make money* or *to make actually money,* both of which are more difficult to read and have the potential for being interpreted incorrectly.]

subject
The noun or pronoun in a sentence about which something is asked or asserted. See *noun, pronoun.* See also Chapters 1, 2, 3.

> The *technician* operated a strobe light during the experimental wing's vibration test.
> [This sentence asserts information about what the technician was doing.]

subjective case
A synonym for the *nominative case.* The subjective case includes subjects and predicate nouns. See *subject, noun, predicate noun.* See also 2.4.

> Magnetically levitated *trains* often use low-temperature superconductors.
> [The subject *trains* is in the subjective or nominative case.]

subjunctive mood
See *mood.*

subordinate clause

Also called a *dependent clause*. A clause that is dependent on another part of the sentence. See *clause, dependent clause*.

She speaks faster *than he can think*.
[In this sentence, the subordinate clause *than he can think* is relative to, and dependent on, how fast she speaks.]

subordinate conjunction

Also called a *subordinating conjunction*. Conjunctions used to join two clauses together while subordinating one of the clauses to the other. See *conjunction*. See also 8.3.

The message was garbled *because* the transfer rate was too high.
[The first clause is connected to the second with *because*. In the process, this conjunction also subordinates the second clause to the first.]

subordinate pronoun. (also subordinating pronoun)

See *relative pronoun*.

subordinator

A conjunction or pronoun that changes the clause that follows from a main clause to a subordinate clause. See *subordinate clause, relative pronoun, subordinate conjunction*. See also 3.2.5, 8.3.

He understands the process.
[independent or main clause]

***Whether* he understands the process**
[*Whether* subordinates this clause.]

syntax

The mechanical arrangement of words, phrases, and clauses to form sentence structure.

tense
The form of a verb that shows its function with respect to the present, past, and future. See 5.2, Table 5.1.

Table 12.6 Tenses of the English language

present simple	present simple progressive
past simple	past simple progressive
future simple	future simple progressive
present perfect	present perfect progressive
past perfect	past perfect progressive
future perfect	future perfect progressive

transitional phrase
A phrase that functions like an adverbial conjunction. See *adverbial conjunction.* See also 6.5, 8.4, Table 6.4.

The experimental aircraft handled well; *more importantly,* it had no system failures.
[The transitional phrase, *more importantly,* is punctuated as an adverbial conjunction.]

transitive verb
A verb that takes an object. See *intransitive verb.* See also 5.5.1.

The technician *operated* the milling machine.
[The verb *operated* takes the direct object *machine* and is transitive.]

two-word verb
A construction that pairs a verb with a preposition, resulting in a combination that frequently has an idiomatic meaning. See *idiom.* See also 7.2, Tables 7.2 and 7.3.

The researcher is *wrapping up* the experiment.
[The two-word verb *wrapping up* does not literally mean enclosing the experiment in paper with a ribbon tied around it; rather, it means *finishing* or *completing* the experiment.]

uncountable noun.
See *mass noun.*

verb
The action word of a predicate, which asks or asserts something about the subject in a sentence. See *active voice, passive voice, tense, number.* See also Chapter 5.

verb stem
The base form of a verb. See *verb.* See also 5.2.

verbal
A verb used as a noun, pronoun, adjective, or adverb. See *gerund, participle, infinitive.* See also 4.1.2.

***Speaking* like a human was easy with voice synthesis technology, but the *talking* robot was not able *to think* like a human.**
[In the sentence above, *speaking* is a gerund acting as the subject, *talking* is a participle acting as an adjective, and *to think* is an infinitive acting as an adverb.]

voice

The construction and use of a transitive verb that determines whether the receiver of the verb's action precedes or follows the verb. See *active voice, passive voice.* See also 5.5.

The satellite's thrusters *inserted* it into orbit.

[Active voice: the receiver of the action, the pronoun *it,* follows the verb *inserted.*]

The satellite *was inserted* into orbit by its thrusters.

[Passive voice: the receiver of the action, the noun *satellite,* precedes the verb *was inserted.*]

Index

153